D0617220

Electrical Wiring: Design and Applications

John T. Earl

Prentice-Hall, Inc., Englewood Cliffs, New Jersey 07632

Library of Congress Cataloging in Publication Data

Earl, John T., 1938-
  Electrical wiring, design and applications.

  Includes index.
  1. Electric wiring, Interior. I. Title
TK3271.E32 1986   621.319′24   85-9298
ISBN 0-13-247685-1

Editorial/production supervision and
    interior design: Nancy Velthaus
Cover design: Joseph Curcio
Manufacturing buyer: Rhett Conklin

© 1986 by Prentice-Hall, Inc., Englewood Cliffs, New Jersey 07632

*All rights reserved. No part of this book may be
reproduced, in any form or by any means,
without permission in writing from the publisher.*

Printed in the United States of America

10 9 8 7 6 5 4 3 2

ISBN 0-13-247685-1 01

Prentice-Hall International, Inc., *London*
Prentice-Hall of Australia Pty., Limited, *Sydney*
Editora Prentice-Hall do Brasil, Ltda., *Rio de Janeiro*
Prentice-Hall Canada Inc., *Toronto*
Prentice-Hall Hispanoamerica, S.A., *Mexico*
Prentice-Hall of India Private Limited, *New Delhi*
Prentice-Hall of Japan, Inc., *Tokyo*
Prentice-Hall of Southeast Asia Pte. Ltd., *Singapore*
Whitehall Books Limited, *Wellington, New Zealand*

Contents

Chapter 23   Farm Wiring   264

Chapter 24   Wiring in Hazardous Locations   276

Chapter 25   Standby and Emergency Power Systems 295

Chapter 26   Basis of Electrical Blueprint Reading   299

Appendix A   Coefficients of Utilization   332

Preface

The electrical construction industry in the United States and other nations continues to grow at a phenomenal rate. In the United States alone, electrical construction will approach $30 billion a year by the time this book is printed. With the continued demand for electrical systems in building construction—both new and existing alike—the need for journeymen electricians, technicians, engineers, and others in the electrical field increases also.

This book is designed to provide up-to-date information on theory, design, and practical applications to help all concerned with the electrical industry to approach their work with more practical solutions to problems, which in turn will instill more confidence in their work. This book will be helpful to all persons working in the electrical construction industry, including vo-tech students, apprentice electricians, journeyman electricians, electrical designers, draftsmen and engineers, architects, electrical contractors, and specification writers.

Although this book is designed mainly for the professional and practicing trade personnel, it is also a book that many homeowners will find of value, especially the do-it-yourselfers who aspire to the finest workmanship in their home repair projects.

Any book of this type cannot really be written by one person; it takes the joint efforts of many—including the manufacturers, the NE Code, the International Brotherhood of Electrical Workers (where I got my start), typists, editors, and many others. To those who helped with this project, my most sincere appreciation is expressed—for without them, there would be no book.

John T. Earl

1

Introduction to Electrical Technology

Since the first central-station electric generating plant was developed in New York City in 1882 by Thomas A. Edison, the electrical construction industry has grown at an astonishing rate to become one of the largest industries in the United States. The first generating plant created public demand for the use of electric lighting and power in existing buildings, as well as for new construction.

These first electrical wiring installations were usually laid out by workers employed and trained by the power companies, and the majority of these installations were "designed" by the mechanics on the job, often as the work progressed. Building contractors then began hiring mechanics of their own to install electrical wiring systems, but because of the special skills and knowledge required, these same builders soon began leaving the wiring installations to mechanics who began to specialize in this work as electrical contractors.

As the electrical construction continued to become a more important part of the general building construction, architects began to prepare layouts of the desired electrical systems on their architectural drawings. This layout usually indicated the lighting outlets, base "plugs," and light switches by means of certain symbols. A line was sometimes drawn from a lighting outlet to a wall switch to indicate how the various lamps were to be controlled, but this was usually the extent of the electrical design. The details of wiring, number of circuits, and the like, were still left to the mechanics installing the system. As electrical systems became more extensive and complex, electrical contractors began hiring draftsmen to prepare working drawings to supplement the sketchy outlet layout on the architectural drawings, to provide a basis for preparing estimates, and to give instructions to electricians in the field who were installing the systems. These early electrical draftsmen were probably experienced architectural or mechanical draftsmen who were trained in the field of electricity by the electrical contractors.

From that point on, electrical construction continued to become a more important part of general building construction, and soon the architects began to prepare more extensive layouts of the electrical systems, until finally separate drawings were included along with the architectural drawings. As the volume of such layout work increased and electrical systems became still more extensive and complex, a greater engineering knowledge of power and illumination requirements became necessary. Persons with the proper knowledge and training began to devote their time exclusively to designing and laying out electrical installations as consulting engineers, selling their services to the architects. These consulting engineers conveyed their designs by means of working drawings that used symbols, lines, notations, and so on. Thus, the electrical designer has become a very important cog in the wheel of electrical construction.

THE NATIONAL ELECTRICAL CODE (NE CODE)

Owing to the potential fire and explosion hazards caused by the improper handling and installation of electrical wiring, certain rules in the selection of materials, quality of workmanship, and precautions for safety must be followed. To standardize and simplify these rules and provide a reliable guide for electrical construction, the National Electrical Code (NE Code) was developed. The NE Code, originally prepared in 1897, is frequently revised to meet changing conditions, improved equipment and materials, and new fire hazards. It is a result of the best efforts of electrical engineers, manufacturers of electrical equipment, insurance underwriters, fire fighters, and other concerned experts throughout the country.

The NE Code is now published by the National Fire Protection Association (NFPA), 470 Atlantic Avenue, Boston, Massachusetts 02210. It contains specific rules and regulations intended to help in the practical safeguarding of persons and property from hazards arising from the use of electricity.

Although the NE Code itself states, "This Code is not intended as a design specification nor an instruction manual for untrained persons," it does provide a sound basis for the study of electrical design and installation procedures—under the proper guidance. The probable reason for the NE Code's self-analysis is that the code also states, "This Code contains provisions considered necessary for safety. Compliance therewith and proper maintenance will result in an installation essentially free from hazard, but not necessarily efficient, convenient, or adequate for good service or future expansion of electrical use."

The NE Code, however, has become the bible of the electrical construction industry, and anyone involved in electrical work, in *any* capacity, should obtain an up-to-date copy, keep it handy at all times, and refer to it frequently.

NE CODE CHAPTERS

In order that those involved in the electrical construction industry may understand the language and terms used in the NE Code, as well as in the industry in general, the definitions listed in Chapter 1, Article 100 of the code should be fully understood. For simplicity, this article lists only definitions essential to the proper use of the code; and then terms used in two or more other articles are defined in full in this article. Other definitions, however, are defined in other individual articles of the NE Code where they apply.

General requirements for electrical installations are given in Article 110. This article, along with Article 100, should be read over several times until the information contained in both is fully understood and firmly implanted in the reader's mind. With a good understanding of this basic material, the remaining portions of the NE Code are easier to understand.

Chapter 2, Wiring Design and Protection, of the NE Code is the chapter that most electrical designers, workers, and others in the field will use the most. It covers such data as using and identifying *grounded conductors, branch circuits, feeders, calculations, services, overcurrent protection,* and *grounding,* all necessary for any type of electrical system, regardless of the building type in which the system is installed.

Chapter 2 of the NE Code deals mainly with how-to items, that is, how to provide proper spacing for conductor supports, how to provide temporary wiring, how to size the proper grounding conductor or electrode, and so forth. If a problem develops pertaining to the design or installation of a conventional electrical system, the answer can usually be found in this chapter.

Rules governing the wiring method and materials used in a specific installation are found in Chapter 3, Wiring Methods and Materials, of the NE Code. The provisions of this chapter apply to all wiring installations except remote-control switching, low-energy power, signal systems, communication systems, and conductors that form an integral part of equipment, such as motors, motor controllers, and the like.

Rules pertaining to raceways, boxes, cabinets, and raceway fittings are also found in Chapter 3 of the NE Code. Since outlet boxes vary in shape to accommodate the size of the raceway, the number of conductors entering the box, the type of building construction, the atmospheric conditions of the building, and other special requirements, Chapter 3 of the NE Code is designed to answer most questions that might arise.

Chapter 3 of the NE Code also generally covers a wide variety of switches, push buttons, pilot lamp receptacles, and convenience outlets, as well as switchboards and panelboards. Such items as location, installation methods, clearances, grounding, and overcurrent protection are thoroughly covered in this section of the code.

Chapter 4, Equipment for General Use, of the NE Code begins with the use and installation of flexible cords and cables, including the trade name, type letter, wire size, number of conductors, conductor insulation, outer covering, and use of each. The articles continue on to fixture wires, again giving trade name, type letter, and other pertinent details.

The article on lighting fixtures in Chapter 4 is especially of interest to electricians and designers as it gives the installation procedures for the various types of fixtures for use in specific locations (fixtures near combustible materials, in closets, etc.).

The selection of electric motors is found in Articles 430 through 445 of the NE Code, as well as mounting the motor and making electrical connections to it. Heating equipment, transformers, and capacitors are also found in Chapter 4 of the NE Code.

While storage batteries are not often thought of as part of an electrical wiring system for building construction, they are often used to provide standby emergency lighting service and to supply power to security systems separate from the main ac electrical system. Most requirements pertaining to battery-operated systems will be found in Chapter 4 of the NE Code.

Any area where the atmosphere or material in the area is such that the sparking of operating electrical equipment may cause an explosion or fire is considered a hazardous location, and these areas are covered in Chapter 5 of the NE Code, Special Occupancies. These locations have been classified in the NE Code into certain class locations, and various atmospheric groups have been established on the basis of the explosive character of the atmosphere for the testing and approval of equipment for use in the various groups.

The basic principle of explosion-proof wiring is to design and install a system so that, when the inevitable arcing occurs within the electrical system, ignition of the surrounding explosive atmosphere is prevented. The basic principles of such an installation are covered in Chapter 5 of the NE Code.

Theaters and similar occupancies also fall under the regulations set forth in Chapter 5 of the NE Code. Recognizing that hazards to life and property due to fire and panic exist in theaters, there are certain requirements in addition to those of the usual commercial wiring installations. While drive-in-type theaters do not present the inherent hazards of enclosed auditoriums, the projection rooms and other areas adjacent to these rooms must be properly ventilated and wired for the protection of operating personnel and others using the areas.

Other areas falling under the regulations of Chapter 5 of the NE Code, Special Occupancies, include residential storage garages, aircraft hangars, service stations, bulk-storage plants, finishing processes, health-care facilities, mobile homes and parks, and recreation vehicles and parks.

The provisions in Chapter 6, Special Equipment, of the NE Code apply to electric signs and outline lighting, cranes and hoists, elevators, electric welders, and sound-recording and similar equipment. Therefore, any electrician or designer who works on any of these systems should thoroughly check through Chapter 6 of the NE Code, along with other chapters that may apply.

Electrical signs and outline lighting are usually considered to be self-contained equipment installed outside and apart from the building wiring system. However, the circuits feeding these lights are usually supplied from within the building itself. Therefore, some means of disconnecting such equipment from the supply circuit is required. All such equipment must be grounded except when insulated from the ground or conducting surfaces or if inaccessible to unauthorized persons.

Neon tubing, where used, requires the use of step-up transformers to provide the necessary operating voltages, and secondary conductors must have insulation and be rated for this high voltage; terminators must also be of the proper type and must be protected from or inaccessible to unqualified persons. In most instances, the electrician will be responsible only for providing the feeder circuit to the location of the lights; qualified sign installers will usually do the actual sign work.

Cranes and hoists are usually furnished and installed by those other than electricians. However, it is usually the electrician's responsibility to furnish all wiring, feeders, and connections for the equipment. Such wiring will consist of the control and operating circuit wiring on the equipment itself and the conductors supplying electric current to the equipment in a manner to allow it to move or operate properly. Furthermore, electricians are normally required to furnish and install the contact conductor and necessary suspension and supporting insulators. Motors, motor-control equipment, and similar items are normally furnished by the crane manufacturer.

The majority of the electrical work involved in the installation and operation of elevators, dumbwaiters, escalators, and moving walks is usually furnished and installed by the manufacturer. The electrician is usually required only to furnish a feeder terminating in a disconnect means in the bottom of the elevator shaft and perhaps to provide a lighting circuit to a junction box midway in the elevator shaft for connection of the elevator-cage lighting cable. Articles in Chapter 6 of the NE Code will give most of the requirements for these installations.

Electric welding equipment is normally treated as a piece of industrial power equipment for which a special power outlet is provided. Certain specific conditions, however, apply to circuits supplying welding equipment and are outlined in Chapter 6 of the NE Code.

Wiring for sound-recording and similar equipment is essentially of the low-voltage type. Special outlet boxes or cabinets are usually provided with the equipment, although some items may be mounted in or on standard outlet boxes. Some systems of this type require direct current, which is obtained from rectifying equipment, batteries, or motor generators. The low-voltage alternating current is obtained through the use of relatively small transformers connected on the primary side to a 120-volt circuit within the building.

Other items covered under Chapter 6 of the NE Code include x-ray equipment, induction and dielectric heat-generating equipment, and machine tools. A brief reading of this chapter will provide a sound basis for approaching such work in a professional manner, and then problems can be studied in more detail during installation.

In most commercial buildings, codes and ordinances require a means of lighting public rooms, halls, stairways, and entrances so that, if the general building lighting is interrupted, there will be sufficient light to allow the occupants to exit from the building. Exit doors must also be clearly indicated by illuminated exit signs.

Articles in Chapter 7, Special Conditions, of the NE Code give provisions for installing emergency lighting systems. Such circuits should be arranged so that they may be automatically transferred over to an alternate source of current supply (from storage batteries, gasoline-driven generators, or properly connected to the supply side of the main service) so that disconnecting the main service switch will not disconnect the emergency circuits. Additional details may be found in Article 700 of the NE Code.

Circuits and equipment operating at more than 600 volts between conductors will also be found in this portion of the code. In general, conductor insulations must be of a type approved for the operating voltage, and the conductors must be installed in rigid conduit, duct, or armored cable approved for the voltage used. These cables must also be terminated with approved cable-terminating devices. All exposed live parts must be given careful attention and adequately guarded by suitable enclosures or isolated by elevating the equipment beyond the reach of unauthorized personnel. Overcurrent protection and disconnecting means must be manufactured specifically for the operating voltage of the system. Examples of such high-voltage installations include feeders for synchronous motors and condensers, substations and transformer volts, and the like. Other details are covered in articles in Chapter 7 of the NE Code.

Among other items in Chapter 7 of the NE Code is the installation of outside wiring other than for electric signs. Such wiring is either attached to the building or between two or more buildings, run overhead, underground, or in a raceway fastened to the face of the building.

Overhead systems may consist of individual conductors supported on or by insulators on or at the building surface, or by means of approved cable either attached to or suspended between buildings. When the buildings are some distance from each other, intermediate supporting poles are necessary, and certain clearance distances must be maintained over driveways and buildings.

Underground wiring between buildings or outside buildings is installed either directly in the ground as direct-burial cable or else pulled through raceways consisting of rigid conduit, PVC conduit, or ducts encased in concrete.

Chapter 8, Communication Systems, of the NE Code deals with communication systems and circuits, that is, telephone, telegraph, district messenger, fire and burglar alarms, and similar central station systems and telephone systems not connected to a central station system but using similar types of equipment, methods of installation, and maintenance. Articles in this chapter also cover radio and television equipment, community antenna television and radio distribution systems (cable TV), and similar systems. The basic requirements are outlined in this chapter of the code, but a good knowledge of communications systems is also required for a proper installation.

Local inspection authorities frequently review actual construction sites to make a determination to grant exceptions for the installation of conductors and equipment not under the exclusive control of the utility or power companies and used to connect the electric utility supply system to the service-entrance conductors of the premises served, provided such installations are outside a building or terminate immediately inside a building wall.

Furthermore, the authority having jurisdiction on a particular project may waive specific requirements in the NE Code or permit alternate methods, where it is assured that equivalent objectives can be achieved by establishing and maintaining effective safety procedures.

The NE Code is also intended to be suitable for mandatory application by government bodies exercising legal jurisdiction over electrical installations and for use by insurance inspectors. The authority having jurisdiction for enforcement of the NE Code will have the responsibility for making interpretations of the rules, for deciding on the approval of equipment and materials, and for granting the special permission contemplated in a number of the rules.

USING THE NATIONAL ELECTRICAL CODE

The drawing in Fig. 1-1 illustrates the NE Code requirements for a complete electrical distribution system from the generating plant through conversion and transmission to distribution and utilization. The use of this illustration is briefly explained as follows.

Assume that an electrician is required to wire a residential building or dwelling. In most occupancies of this type, the local power company will furnish electric service to a point on the building. The owner is then required to furnish a service entrance (cable or conduit) to the watt-hour meter base

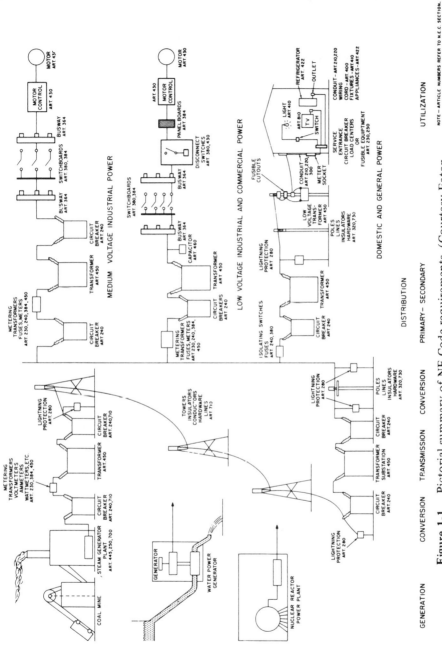

Figure 1-1 Pictorial summary of NE Code requirements. (Courtesy Eaton Corporation, Cutler-Hammer Products.)

and from here to a disconnect or panelboard inside the building. Furthermore, the owner will be responsible for all interior wiring, that is, power and lighting outlets, branch circuit wiring, feeders for heavy appliances, and the like.

In wiring the building, the electricians will usually lay out the various outlets first and then size the branch circuits and service entrance. By "laying out," we mean to locate and mark where each outlet is to be placed in the building structure. This may be first done in the form of a sketch or working drawing or measured and laid out directly in the building. Note in Fig. 1-1 (in the lower right corner under "Domestic and General Power") that Articles 210, 220, and 300 of the NE Code are given as reference at the service-entrance conduit.

Article 210, Branch Circuits, covers the general layout and requirements of all branch circuits within the building, including the spacing of duplex receptacles in residential occupancies. Knowledge of these requirements is necessary to properly calculate the size of the service entrance.

Article 220 gives the procedure for calculating the size of the service entrance, as well as branch circuits and feeders. Finally, Article 300 will inform the electrician as to the correct wire size and the type of insulation required.

Once inside the building, Articles 230 and 250 of the NE Code (as shown in Fig. 1-1) outline the requirements for circuit breakers or fusible service equipment and load centers. Again, Articles 210 and 220 are used for reference when sizing and installing the branch circuits. Article 410 covers lighting fixtures and Article 422 covers appliances. Article 800 deals with TV antennas and outlets and other communication systems.

Under "Low Voltage Industrial and Commercial Power" in Fig. 1-1, additional articles are given for switchboards, busways, capacitors, transformers, and the like. Until the contents of the NE Code become second nature, the student should keep this diagram handy for reference when working on any type of building or electrical system. Also use it when checking over blueprints or other forms of construction documents. Although this diagram is merely a rough outline of all the possibilities, it does form a good basis for quickly locating certain articles in the NE Code.

LOCAL CODES AND ORDINANCES

A number of towns, cities, and counties have their own local electrical code or ordinance. In general, these are based on, or similar to, the NE Code, but on certain classes of work, they may have a few specific rules that are usually more rigid than the NE Code.

In addition to the NE Code and local ordinances of certain cities, local power companies may have some special rules regarding location of service-entrance wires, watt-hour meter connections, and similar details that must be satisfied before connection can be made to a building.

Mandatory rules of the NE Code are characterized by the use of the word *shall,* while advisory rules are characterized by the use of the word *should.* When statements are made using the latter word, they are stated as recommendations, or that which is advised but not necessarily required. Some local ordinances, however, may change the meaning of "should" to "shall." When working in a new area, it is therefore useful to find out if there are legal requirements amending the NE Code and, specifically, what these are.

REQUIREMENTS OF ARCHITECTS AND ENGINEERS

When any type of building of any consequence is contemplated, an architect is usually commissioned to prepare the complete working drawings and written specifications for the building. For jobs of any consequence, the architect usually includes drawings and specifications for the complete electrical system, which are usually prepared by consulting engineering firms.

Consulting engineers will often specify materials and methods that surpass the requirements of the NE Code in order to obtain a higher-quality finished job. Electricians who work on such jobs must comply with the working drawings to carry out the engineer's design. For example, number 14 AWG wire may be quite adequate for a certain wiring installation according to the NE Code. However, the engineer may specify that wire smaller than number 12 AWG cannot be used on this particular project. If so, the installation must be carried out as specified, even though it surpasses the NE Code. For such reasons, all persons involved in a building construction project should carefully study the working drawings and construction drawings before starting the wiring installation and refer to them often as the work progresses.

Reading electrical drawings will be covered in a later chapter in this book. You may also want to obtain a copy of *Electrical Blueprint Reading,* published by Craftsman Book Company, 6058 Corte Del Cedro, Carlsbad, California 92008.

TESTING LABORATORIES

There are several qualified testing laboratories in the United States and Canada, but Underwriters Laboratories (UL), Inc., is the most widely used. It investigates, studies, experiments, and tests products, materials, and systems. If the items are found to meet the UL safety requirements during these tests, UL will list the items. Bear in mind, however, that UL does not approve such items; it only lists them as having passed their tests for safety. Therefore, the term "listed by UL" is the correct one rather than "approved by UL."

Figure 1-2   All equipment tested by Underwriters' Laboratories bears a seal similar to the one shown here. (Courtesy Underwriters' Laboratories, Inc.)

Tested products that meet UL standards are listed under various categories in directories published by UL. They can be purchased from Underwriters Laboratories, Inc., 333 Pfingsten Road, Northbrook, Illinois 60062.

The products and/or the containers in which the products are shipped should contain the UL listing mark as shown in Fig. 1-2. The listing mark may be on an interior or exterior surface of the product, and some will have the mark on the shipping carton or reel.

The electrician should become familiar with this mark and keep in mind that many electrical inspectors and consulting engineers will allow only UL-listed items to be installed in an electrical system. The failure to do so could cost someone considerable time and expense in replacing the nonlisted items.

2

Math for Electrical Technology

The amount of mathematics that you already know is probably quite sufficient for work in electrical technology. However, you may not have used your mathematical knowledge to its fullest extent since it was learned, and anything not put to use is soon forgotten, until it is relearned and put to use. This is the purpose of this chapter: to review mathematical techniques as applied to the electrical industry.

NUMBERS

The system of numbers most often used in everyday life is called the *decimal system.* Including zero, this number system contains ten digits: 1, 2, 3, 4, 5, 6, 7, 8, 9, and 0.

The value of a digit depends not only on the digit itself, but also on its position. For example, take the number 555. The digit on the extreme right has a value of five, but the digit in the middle position has a value of fifty, while the digit in the left position has a value of five hundred. Notice that as the digits progress from right to left, each position of the digit increases its value by ten times the value of the preceding position. To illustrate, the digit 5 in the middle position has a value ten times as great as the digit 5 in the right-hand position. Similarly, the digit 5 in the left-hand position has a value ten times as great as the digit 5 in the middle position.

In any three-place number, such as 555, each digit is given a special position name. Again reading from right to left, the position name of the first digit is *units;* the position name of the second digit is *tens;* and the position name of the third digit is *hundreds.* This simply means that the value of the first digit (the one in the right-hand position) is equal to the digit itself; the value of the second digit is ten times as great as the digit itself (10 × 5 = 50), and the value of the third digit is one hundred times as great as the digit itself.

In writing and reading numbers, the digits are separated into groups of three figures, called *periods.* The number 555, therefore, makes up a first period called *units.* In a number such as 555,555, two periods are present, the second of which is called *thousands.* Each period contains the same three positions: that is, units, tens, and hundreds. The periods may be continued to the left to indicate millions, billions, or even a fifth period to indicate trillions.

Using a number, say, 924,281,735,146 and beginning at the very left of the number, we have "nine hundred twenty-four billion in the fourth period, two hundred eighty-one million in the third period, seven hundred thirty-five thousand in the second period, and one hundred forty-six in the first period. The number is read "nine hundred twenty-four billion, two hundred eighty-one million, seven hundred thirty-five thousand, one hundred forty-six." Notice that the word "and" is not used to connect any of the periods.

BASIC MATH OPERATIONS

The basic math operations are addition, subtraction, multiplication, and division. The electrical technician should be able to use these operations in solving whole number problems, decimal number problems, and problems dealing with fractions. With these four basic math operations, along with squares, square roots, percents, and solving of equations, the electrical technician will have all the basic tools necessary for most electrical calculations that will be encountered.

WHOLE NUMBERS

Addition

In adding whole numbers, the digit on the right of each number is lined up before the columns are added. For example, to add 570 ohms and 23 ohms,

$$
\begin{array}{r}
570 \text{ ohms} \\
+\ \ 23 \text{ ohms} \\
\hline
593 \text{ ohms}
\end{array}
$$

Subtraction

As in addition, the numbers are lined up with the digit on the right of each number. Then proceed to subtract. To subtract, say, 250 volts from 1000 volts,

$$
\begin{array}{r}
1000 \text{ volts} \\
- \ 250 \text{ volts} \\
\hline
750 \text{ volts}
\end{array}
$$

Multiplication

In multiplying whole numbers, line up the digit on the right of the numbers as in the following example,

$$
\begin{array}{r}
250 \text{ feet per roll of wire} \\
\times \ 16 \text{ rolls of wire} \\
\hline
\end{array}
$$

First multiply the digit 6 in the number 16 times 0 in the top number 250; 6×0 is zero. Continue from right to left, multiplying each digit in the top number by the 6 in the lower number. Six (6) times 5 is 30, but since this number has two digits (3 and 0), the zero is recorded (written down), and the digit 3 is placed in the next column to the left, to be added to the product of 6×2, which is 12; then add the 3, which gives 15. The problem thus far should look like the following:

$$
\begin{array}{r}
^{3}250 \\
\times \ 16 \\
\hline
1500
\end{array}
$$

Now it is time to use the digit 1 in the number 16; that is, 1×250. As this multiplication takes place, write the answer under the first answer (1500), using the second column to start the first digit. So 1 times 0 is zero. Place this answer (0) under the second zero from the right in the number 1500. Continuing, $1 \times 5 = 5$ and $1 \times 2 = 2$. The problem should now appear as follows:

$$
\begin{array}{r}
250 \\
\times \ 16 \\
\hline
1500 \\
+250 \\
\hline
\end{array}
$$

Now add the two resulting numbers, and the answer gives the total feet of wire in 16 rolls if each roll contains 250 feet.

$$
\begin{array}{r}
250 \ \text{feet per roll of wire} \\
\times \quad 16 \ \text{rolls of wire} \\
\hline
1500 \\
+250 \\
\hline
4000 \ \text{total feet of wire}
\end{array}
$$

Multiplication may be written 24 X 15, (24)(15), 24 · 15, or 24 times 15. In the example (24 · 15), do not confuse the dot between the numbers with a decimal. Because of this potential confusion, this method of showing multiplication is rarely used.

Division

Division of whole numbers is accomplished as follows: Assume that a 252-foot coil of wire must be divided into 12 equal lengths. The problem would begin as follows:

$$12 \overline{)252}$$

How many times will 12 go into 2? It will not go, so a 0 is placed above the 2:

$$
\overset{0}{12 \overline{)252}}
$$

Now ask how many times 12 goes into 25. Since 2 X 12 = 24, the digit 2 is placed above 5 in the number 25:

$$
\overset{02}{12 \overline{)252}}
$$

This digit 2 is then multiplied by the 12 to obtain an answer of 24, which is then placed under the first two digits of the number 252, or below the 25. The next step is to subtract 24 from 25, which leaves a remainder of 1, as shown.

$$
\begin{array}{r}
02 \\
12 \overline{)252} \\
-24 \\
\hline
1
\end{array}
$$

The remaining digit (2) in the number 252 is brought down and placed alongside the 1 obtained from subtracting.

$$
\begin{array}{r}
02 \\
12\,)\overline{252} \\
-24\!\downarrow \\
\hline
12
\end{array}
$$

This operation provides the number 12. Then 12 into 12 is exactly 1. Multiplying this 1 by 12 gives 12, and this 12 subtracted from the first 12 leaves zero. The number obtained in the previous operation is called the *remainder*.

$$
\begin{array}{r}
021 \\
23\,)\overline{252} \\
-24 \\
\hline
12 \\
12 \\
\hline
0 \text{ remainder}
\end{array}
$$

Therefore, 252 feet of wire divided by 12 gives 12 lengths of 21 feet each. Division may be written

40 divided by 8, $\dfrac{40}{8}$, 40/8, 40 : 8, 40 ÷ 8, $8\,)\overline{40}$

DECIMALS

The number 22-1/2 can be written 22.5 and is known as a decimal number. The number is read "twenty-two and five-tenths," or "twenty-two decimal five." In the field you will also hear the number called "twenty-two point five," but the most accepted way is to call the number "twenty-two and five-tenths."

Take the number 29.271642. After the whole number (29), the digit places to the right of the decimal are as follows:

2 9 . 2 7 1 6 4 2

tenths
hundredths
thousandths
ten thousandths
hundred thousandths
millionths

Therefore, the number 29.271 is read "twenty-nine and two hundred seventy-one thousandths." If a number does not have a decimal in it, such as, say, 29, take the decimal to be to the right of the number, thus, 29., and it is the same as the whole number 29 used previously.

Addition of Decimal Fractions

To solve the problem of 22.35 volts + 0.241 volt + 3.9 volts, the first step is to line up the decimals in the problem and the answer in a straight line, with the numbers in their proper places on both sides of the decimal. Now add each column just as if this were a whole-number problem.

$$
\begin{array}{ll}
\downarrow & \\
22.35 & \text{volts} \\
.241 & \text{volt} \\
\underline{3.9} & \text{volts} \\
26.491 & \text{volts} \\
\uparrow &
\end{array}
$$

Subtraction of Decimal Fractions

To subtract decimal fractions, as in addition, the decimals in the problem and answer must be in line with the numbers in the proper places on either side. Then subtract as if the numbers were whole numbers. For example, subtract 12.3 feet from 72.54 feet.

$$
\begin{array}{ll}
\downarrow & \\
72.54 & \text{feet} \\
\underline{12.3} & \text{feet} \\
60.24 & \text{feet} \\
\uparrow &
\end{array}
$$

Multiplication of Decimal Fractions

When solving multiplication problems that contain decimal fractions, first write down the problem and solve it as if there were no decimals. Multiply 25.63 times 4.2.

$$
\begin{array}{r}
25.63 \\
\times \quad 4.2 \\
\hline
5126 \\
10252 \quad \\
\hline
107646
\end{array}
$$

Now the decimal has to be placed in the answer. To do this, count the decimal places to the right of the decimal in the problem. Looking at the numbers 25.63 and 4.2, it is found that there are two decimal places to the right of 25 and one decimal place to the right of 4. Therefore, count over three decimal places from the right in the answer and put in the decimal. Thus, the answer to the problem is 107.646.

Division of Decimal Fractions

A division decimal problem is solved by first moving the decimal in the *divisor* (the number you are dividing by) as many spaces to the right as necessary to make it a whole number. Then move the decimal in the *dividend* (the number being divided) the same number of places. Divide 5.492 feet by 2.5 feet.

$$2.5\overline{)5.492} = 25\overline{)54.92}$$

It was necessary to move the decimal one place to the right to make 2.5 a whole number (25), and therefore the other number had to be treated the same. Now divide and obtain an answer as if there were no decimals in the problem. Thus,

$$
\begin{array}{r}
2.1968 \\
25\overline{)54.9200} \\
50 \\
\hline
49 \\
25 \\
\hline
242 \\
225 \\
\hline
170 \\
150 \\
\hline
200 \\
200 \\
\hline
\end{array}
$$

As many zeros may be added to the right of the decimal as you wish; in this problem, two were added.

FRACTIONS

Fractions result when one whole number is divided by another whole number, such as 3/4. The top number, 3, is called the *numerator,* and the bottom number, 4, is called the *denominator.* Fractions come in three types: regular, 3/4; improper, 5/4; and mixed number, 2-1/4.

Addition of Fractions

Regular, improper, and mixed numbers can all be added by following a few simple steps. Suppose 2-3/5 was to be added to 1-1/2. First, the denominators of the fractions must become the same number (called *common denominator*). A number must be chosen that both 5 and 2 will divide into. Ten is the smallest number that 5 and 2 will both go into. There are other larger numbers, such as 20 and 40, that will work as common denominators, but the smallest is always chosen. If difficulties arise in finding a common denominator, multiply all denominators together and it will produce a number that will always work.

$$2\frac{3}{5} = 2\frac{}{10}$$
$$+1\frac{1}{2} = 1\frac{}{10}$$

To determine what number is to be used for the numerator of each new fraction, multiply the numerator in the problem by the same number that the denominator is multiplied by to form the new denominator.

$$2\frac{3 \times 2}{5 \times 2} = 2\frac{6}{10}$$
$$1\frac{1 \times 5}{2 \times 5} = 1\frac{5}{10}$$

Next, add the whole numbers together and add the numerators together, leaving the denominators the same.

$$2\frac{3}{5} = 2\frac{6}{10}$$
$$+1\frac{1}{2} = 1\frac{5}{10}$$
$$3\frac{11}{10}$$

If the answer is an *improper fraction* (top number larger than the bottom number), divide the bottom number into the top number, in this case 10 into 11, with the result being 1-1/10. Add the 1 to the whole number 3 in the answer, and we get 4-1/10 as the answer to the problem.

Subtraction of Fractions

Subtraction of fractions is solved in a similar way as the addition of fractions. First, obtain a common denominator.

$$4\frac{4}{5} = 4\frac{8}{10}$$
$$-2\frac{1}{10} = 2\frac{1}{10}$$
$$2\frac{7}{10}$$

Thus, the answer to this problem is 2-7/10. However, if the top fraction is smaller than the bottom fraction, another method must be used to solve the problem. For example,

$$5\frac{1}{3} = 5\frac{4}{12}$$
$$-2\frac{3}{4} = 2\frac{9}{12}$$

It becomes obvious that 9/12 cannot be taken away from 4/12. A 1 must be borrowed from the whole number and made into a fraction. Thus, 5 and 4/12 becomes 4 and 1, or 12/12, and 4/12. More simply, 4 + 12/12 + 4/12 = 4-16/12. The problem is then solved thusly:

$$4\frac{16}{12}$$
$$-2\frac{9}{12}$$
$$2\frac{7}{12}$$

Multiplication of Fractions

The electrical technician may recall that multiplication of fractions is usually the easiest operation with fractions. For example, to solve the problem 3/4 × 2/3, multiply the numerators together and multiply the denominators together.

$$\frac{3}{4} \times \frac{2}{3} = \frac{6}{12}$$

Reduce your answer to the smallest fraction possible. This is achieved by dividing the numerator and denominator by the same number that achieves this goal. In this case the number is 6, and the lowest fraction or answer to the problem is 1/2.

To multiply 3-1/2 by 1/5, first change 3-1/2 to an improper fraction. To do this, multiply the whole number (3) by the denominator (2) and add the numerator (1). Thus 3-1/2 becomes 7/2. Then solve the problem.

$$3\frac{1}{2} \times \frac{1}{5} = \frac{7}{2} \times \frac{1}{5} = \frac{7}{10}$$

Division of Fractions

Division of fractions is accomplished in a similar manner to multiplication of fractions. However, the one rule to remember in the division of fractions is to invert (turn upside down) the divisor (number dividing by) and then multiply. For example, solve the problem $3/4 \div 2/3$. First, invert the divisor $2/3$, which becomes $3/2$, and then multiply.

$$\frac{3}{4} \div \frac{2}{3} = \frac{3}{4} \times \frac{3}{2} = \frac{9}{8} \quad \text{or} \quad 1\frac{1}{8}$$

PERCENT

Percent problems can be solved very easily if the problem is turned word for word into a math equation. The math symbols that replace words are the following:

Word	Symbol
What	N
is	$=$
%	$\overline{100}$
of	$\times$

Thus the percent problem

What is 5% of 800?

becomes $\quad N = \dfrac{5}{100} \times 800$

or $\quad N = \dfrac{5}{100} \times 800$

Once a percent problem has been turned into an equation, it falls into one of two types. One type is when the unknown N is by itself on one side of an equation. Refer again to the previous problem. What is 5% of 800? Again substituting symbols for the words,

$$N = \frac{5}{100} \times 800$$

This type is solved simply by multiplying the fractions on the right.

$$N = \frac{5}{100} \times \frac{800}{1}$$

$$N = \frac{4000}{100}$$

$$N = 40$$

The second type is when the unknown N has other numbers with it on the same side of the equation.

What % of 500 is 250?

$$\frac{N}{100} \times 500 = 250$$

When the equation is of this type, apply an "undoing" process to the problem. Undoing is the process of doing just the opposite of what is done to the problem. Untying shoestrings is an undoing process that is the reverse of tying. In this problem, $N/100 \times 500 = 250$, N is divided by 100 on the left side of the $\times$, and N is multiplied by 500 on the right side of the $\times$. Thus, to undo the problem, do just the opposite to both sides of the equation.

$$\boxed{ {\rightarrow}(100)\,\frac{N}{100} \times 500 = 250(100){\leftarrow} }$$
$$\text{Multiply both sides by 100}$$

$$(100)\,\frac{N}{100} \times 500 = 250(100)$$

$$500N = 25{,}000$$

Now do the opposite to both sides, which is to divide both sides by 500.

$$\frac{500N}{500} = \frac{25{,}000}{500}$$

$$N = 50$$

Thus, the answer is 50%.

PROPORTIONS

Two ratios that are equal are known as a proportion. Proportions are one of the easiest types of problems to solve and are probably the most widely used type of problem.

Example 1

$$\frac{3\text{ ohms}}{10\text{ feet of wire}} = \frac{N\text{ ohms}}{50\text{ feet of wire}}$$

$$\frac{3}{10}\diagdown\diagup\frac{N}{50}$$

Multiply diagonally:

$$10N = 3 \times 50$$

$$\frac{10N}{10} = \frac{150}{10}$$

$$N = 15 \text{ ohms}$$

Example 2

$$\frac{500 \text{ feet of wire}}{80 \text{ receptacles}} = \frac{100 \text{ feet of wire}}{N \text{ receptacles}}$$

$$\frac{500}{80} \diagdown \frac{100}{N}$$

$$500N = 80(100)$$

$$\frac{500N}{500} = \frac{8000}{500}$$

$$N = 16 \text{ receptacles}$$

POWERS AND ROOTS

Powers and roots of numbers are best acquired by entering the problem to be solved on a small calculator that has these functions. If one does not have access to such calculators, the powers may be calculated by repeat multiplication,

$$2^5 \quad \text{means} \quad 2 \times 2 \times 2 \times 2 \times 2$$

The answer is 32.

Roots may be found using a chart, a calculator, trial-and-error methods, and other methods. The most accurate and easiest is to use a calculator.

ELECTRICAL EQUATIONS

The electrical technician can solve most electrical equations if he or she isolates the unknown quantity on one side of an equation by using the undoing method to move numbers from the side with the unknown quantity. For example, see Fig. 2-1.

$$R_1 = 20r, \qquad R_2 = 30r, \qquad \text{total resistance} = 10r$$

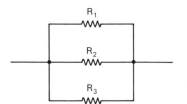

R_1

R_2

R_3 **Figure 2-1**

To find R_3, place the values in the parallel resistor equation.

$$\frac{1}{R_t} = \frac{1}{R_1} + \frac{1}{R_2} + \frac{1}{R_3}$$

$$\frac{1}{10} = \frac{1}{20} + \frac{1}{30} + \frac{1}{R_3}$$

The fractions 1/20 and 1/30 are added to the side that R_3 is on, so subtract them from both sides as an undoing procedure.

$$\frac{1}{10} - \frac{1}{20} - \frac{1}{30} = \frac{1}{20} + \frac{1}{30} + \frac{1}{R_3} - \frac{1}{20} - \frac{1}{30}$$

$$\frac{1}{10} - \frac{1}{20} - \frac{1}{30} = \frac{1}{R_3}$$

Find common denominators:

$$\frac{6}{60} - \frac{3}{60} - \frac{2}{60} = \frac{1}{R_3}$$

$$\frac{6}{60} - \frac{5}{60} = \frac{1}{R_3}$$

$$\frac{1}{60} = \frac{1}{R_3}$$

Since this is a proportion, multiply diagonally.

$$1 \times R_3 = (60)(1)$$

$$R_3 = 60 \text{ ohms}$$

As long as the value of all variables is known except one variable and you have the equation that relates the variables, you should be able to use equation-solving techniques such as the undoing process to solve such equations.

3

Electric Circuits

ELECTRICAL UNITS

The three basic electrical units in any electrical circuit are the ampere (A), volt (V), and ohm (Ω). The current flow in a circuit is measured in amperes. The electromotive force that causes the current to flow is measured in volts. The unit of opposition to current flow is measured in ohms.

Ampere

The rate at which electricity (current) flows through a conductor is represented by the unit called the ampere and may be compared to the rate of flow of water through a pipe in gallons per second. For all practical purposes, the unit strength of an ampere is represented when an electrical current passing through a specified solution of nitrate of silver in water deposits silver at the rate of 0.001118 gram per second. If twice that amount of silver is deposited during one second, the current is 2 amperes, and so on.

Volt

To overcome the resistance of conductors and cause current to flow, an external force is necessary. This force is commonly called *voltage* since the unit of measurement is the volt. This force is also referred to as electromotive force or electric pressure. The electromotive force that will cause a current of 1 ampere to flow through a resistance of 1 ohm is equal to 1 volt. A kilovolt (kV) = 1000 volts, a millivolt (mV) = 0.001 volt, and a microvolt (μV) = 0.000001 volt.

Ohm

All substances offer resistance to the flow of electricity (current) through them. This opposition, or resistance, is measured with a unit called the ohm. The resistance of all metals increases with increase in temperature, while the resistance of carbon, insulating materials, and electrolytic solutions decreases with an increase in their temperature.

The units just described would have no relationship with each other unless they were defined. A very important basic relationship among these units has been established and is called *Ohm's law.* This law states that current flow is proportional to voltage but is inversely proportional to resistance. In other words, the law states that the current, in amperes, increases and decreases directly with an increase or decrease of the pressure difference in volts. It further states that when the resistance is doubled, only half as much current will flow (when the voltage remains the same), and when half the resistance is present, twice as much current will flow. See Fig. 3-1.

The basic ways of stating Ohm's law when I = amperes, R = resistance in ohms, and E = volts are as follows:

1. $E = IR$, or the voltage is equal to the current multiplied by the resistance.
2. $I = E/R$, or the current equals the voltage divided by the resistance.
3. $R = E/I$, or the resistance equals the voltage divided by the current in amperes.

The use of these equations enables us to calculate the third quantity if any two are already known.

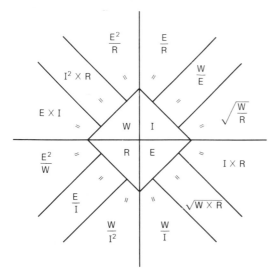

Figure 3-1   Summary of Ohm's law.

Watt

Another important unit of electrical measurement is the watt, the unit of power. Power is defined as the rate at which work is done or the rate at which energy is expended. The watt may also be incorporated into Ohm's law for further calculations. When W = watts, current may be found by the following equations:

1. $I = W/E$, or the current equals the wattage divided by the voltage.
2. $I = \sqrt{W}/R$, or the current equals the square root of the wattage divided by the resistance.

Voltage may be found by using the following equations:

1. $E = W/I$, or the voltage equals the wattage divided by the current.
2. $E = \sqrt{W} \times R$, or the voltage equals the square root of the wattage times the resistance.

Resistance may be found by the following equations:

1. $R = E^2/W$, or the resistance equals the voltage squared divided by the wattage.
2. $R = W/I^2$, or the resistance equals the wattage divided by the current squared.

The power in watts (wattage) of a circuit may be found by the following equations:

1. $W = E^2/R$, or the wattage equals the voltage squared divided by the resistance.
2. $W = I^2 \times R$, or the wattage equals the current squared times the resistance.
3. $W = E \times I$, or the wattage equals the voltage times the current.

CIRCUITS

For an electrical circuit to be complete, it must provide a path for the electric current to flow. All electric circuits consist of one of two distinct types or a combination of the two, that is, series and parallel.

In the series circuit (Fig. 3-2), all components are connected in tandem; it is used very often in control circuits, for conventional magnetic contactor controls, static controls, and electronic controls. The following four rules state the conditions that exist in a series circuit:

- The current is the same in all parts of a series circuit.
- The total resistance in a series circuit is equal to the sum of the individual resistances.
- The total voltage applied to a series circuit divides between the resistors in direct proportion to their resistance.
- The sum of the voltage drops across all the resistors in a series resistive circuit is equal to the applied (source) voltage.

A simple parallel circuit is shown in Fig. 3-3. Here the electrical components are connected *across* the lines rather than in tandem with the lines. Most of the circuits encountered on electrical construction work will consist of parallel circuits or a combination of series and parallel circuits.

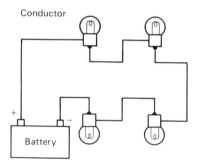

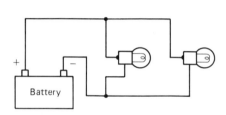

Figure 3-2   Four lamps connected in series.

Figure 3-3   Two lamps connected in parallel.

There are several ways to calculate the total resistance of a parallel circuit, but remember that the total resistance of a parallel circuit is always smaller than the smallest resistor. This is because more paths for the current to flow along are available in a parallel circuit than in a series circuit. Therefore, there is more opportunity for the current to flow, and the total resistance of the circuit becomes lower.

The following are the three most commonly used equations for resistors in parallel:

1. $R_t = \dfrac{R_1}{N}$

2. $R_t = \dfrac{R_1 \times R_2}{R_1 + R_2}$

3. $\dfrac{1}{R_t} = \dfrac{1}{R_1} + \dfrac{1}{R_2} + \dfrac{1}{R_3} + \dfrac{1}{R_4}$

In dealing with the current, voltage, and resistance in a parallel circuit, a few simple rules must be learned:

- The voltage is the same in all parts of a parallel circuit.
- The total current in a parallel circuit is the sum of the currents through the separate parts.

To solve problems for parallel circuits, first draw a circuit diagram of the problem, set down the equation to be used, fill in the equation with the given quantities, and then solve for the missing quantity. A systematic approach to any problem leads to a better understanding of the problem and the theory behind it.

The circuit in Fig. 3-4 shows the resistors R_2 and R_3 connected in parallel, but resistor R_1 is in series with both the battery and the parallel combination of R_2 and R_3. That is, the current flow (indicated by the arrows) leaves the negative terminal of the battery and travels through resistor R_1 and then divides at point A into I_1 and I_2.

The total resistance of this circuit is the sum of R_1 and the resistance of R_2 and R_3 in parallel. Therefore, to find R_t we first need to find the resistance of R_2 and R_3 in parallel. Because these two resistors have identical values, we have a resultant parallel resistance R_p of

$$R_p = \frac{R}{n} = \frac{20 \text{ ohms}}{2 \text{ resistors}} = 10 \text{ ohms}$$

The circuit now looks like Fig. 3-5, in which R_2 and R_3 have been replaced with R_p. We now have a simple series circuit in which the total circuit resistance R_t is

$$R_t = R_1 + R_p = 5 + 10 = 15 \text{ ohms}$$

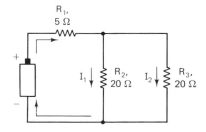

Figure 3-4 A combination series-parallel circuit.

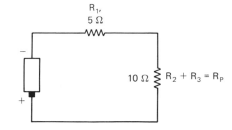

Figure 3-5 The circuit in Fig. 3-4 after combining the resistors R_2 and R_3.

The total circuit current I_t supplied by the battery is then

$$I_t = \frac{E}{R_t} = \frac{30 \text{ volts}}{15 \text{ ohms}} = 2 \text{ amperes}$$

Because R_1 is in series with the battery, we know that the current through R_1 must be 2 amperes. Since R_2 is equal to R_3, equal currents of 1 ampere must flow through each of these two resistances.

Remember always to break up the circuit and work the series or parallel groups separately within the combination; then solve for the whole combination.

When resistors are connected in series, the total resistance can be determined by the equation as previously discussed. Then according to Ohm's law, the total current for the entire circuit can be determined for any given applied voltage. Since this original circuit has all its resistors in series, this current must go through every resistor and back to the voltage source.

An electric direct-current (dc) power supply that produces a voltage will have one terminal of positive polarity and the other of negative. Conventional current flow assumes that current flows from the positive terminal of the supply and returns to the negative terminal. These polarities are easily found with a voltmeter since all dc measuring instruments have polarity indications on their terminals.

When resistances are the only devices used in a circuit, the electrical calculations and observations are the same whether a direct or an alternating current (ac) flows through them. The positive terminal of a dc supply will always be positive. Any one terminal of an ac supply will be alternately positive and then negative. This change of polarity will be uniform among all resistive elements in a purely resistive circuit. Therefore, we can assume a fixed polarity and a constant current direction for the purpose of circuit analysis. This will be valid as long as we are consistent.

Kirchhoff's voltage law states that the net voltage in any continuous path of current must always be zero. When two or more resistors are connected in series, there will be a voltage drop across each resistor. The sum of these voltage drops then must equal the voltage rise of the power supply. The polarity of the voltage drops in a series circuit also can be determined with a voltmeter. According to one of the Ohm's law equations, a voltage drop across any resistor will be

$$\text{Voltage drop} = IR \text{ (see Fig. 3-6)}$$

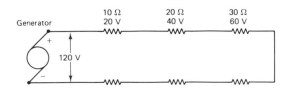

Figure 3-6   A 120-volt circuit with three resistors connected in series showing the voltage drop across each resistor.

One convenient method of writing a voltage equation for a circuit according to Kirchhoff's voltage law is to follow three simple rules:

1. Indicate the current direction with an arrow out of the positive terminal of power supply. If the positive terminal is unknown, assume one as positive.
2. As current goes through a resistor, place a plus sign where current enters a resistor and a minus sign where it leaves.
3. Start at any point in the circuit, in the direction of current flow, and add up all voltages. When going from a+ to a—, this would be taken as a voltage drop and given a minus sign. When going from a— to a+, this would be a voltage rise by a power supply and given a plus sign.

Kirchhoff's voltage law states that the current flow, as permitted by all resistances in the circuit, will cause the voltage drops to be equal to the voltage supply, or

$$E = V_1 + V_2 + V_3 + V_4$$

When resistors or any electrical devices are connected in parallel, there will be different paths for current to flow in. In examining any junction, regardless of the number of current paths, Kirchhoff's current law states that the current of any junction must equal zero. In other words, the current that flows into a junction must equal the current emanating from that junction. Stated another way, the current will flow instantly or not at all, as shown in the following equation:

$$I_T \quad (I_1 + I_2 + I_3 \mid \cdots \mid I_N) = 0$$

When using Kirchhoff's current law, it is necessary to assign arrows to all currents that enter or leave a junction. When assigning these arrows, we assume the current flows from — to + through the power supply and from + to — through an electrical load. The sum of these currents as designated by the arrows must be zero.

MAGNETISM

Anyone working in the field of electricity must be familiar with the principles of magnetism, because generators, transformers, and motors depend on magnets and magnetism for their operation. Even fluorescent lights depend on magnetism for their operation.

A magnet is either permanent or temporary. If a piece of iron or steel is magnetized and retains its magnetism, it is a permanent magnet. A compass is one form of permanent magnet. Others with which you are probably familiar are horseshoe-shaped magnets and bar magnets. Each of these magnets has a north magnetic pole and a south magnetic pole, as do all magnets.

When current flows through a coil, a magnetic field with a north and south pole is set up just like that of a permanent magnet. However, when the current stops, the magnetic field also disappears. This type of temporary magnetism is called *electromagnetism.* Permanent magnets are used for the magnetic field necessary in the operation of small, inexpensive electrical motors. Larger motors, relays, and transformers rely on the magnetic fields from electrical current passing through coils of wire.

When electricity flows through a wire or conductor, magnetic lines of force (magnetic flux) are created around that wire (Fig. 3-7). When a piece of wire is passed through a magnetic field (magnetic lines of force), electricity is created in that wire. We then can readily see the relation between electricity and magnetism. In fact, the very existence of the electrical industry is dependent on magnetism and magnetic circuits.

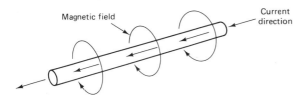

Figure 3-7   When current flows through a conductor, magnetic lines of force are created around the conductor.

RULES FOR DIRECTION OF CURRENT AND MOTION

If the current in a conductor is flowing from south to north and a compass is placed under the conductor, the north end of the needle will be deflected to the west; if the compass is placed over the conductor, the north end of the needle will be deflected to the east (Fig. 3-8). Here are four basic rules to follow when working with magnetism:

1. *To determine the polarity of an electromagnetic solenoid:* In looking at the end of a solenoid, if an electric current flows in it clockwise, the end next to the observer is a south pole and the other end is a north pole; if the current flows counterclockwise, the position of the poles is reversed.

2. *To determine the direction of the lines of force set up around a conductor:* If the current in a conductor is flowing away from the observer, the direction of the lines of force will be clockwise around the conductor.

3. *To determine the direction of motion of a conductor carrying a current when placed in a magnetic field:* Place the thumb, forefinger, and middle finger of the left hand at right angles to each other; if the forefinger shows the direction of the lines of force and the middle finger shows

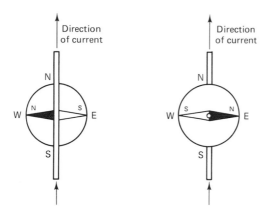

Figure 3-8 If the current in a conductor is flowing from south to north, and a compass is placed under the conductor, the north end of the needle will be deflected to the west; if the compass is placed over the conductor, the north end of the needle will be deflected to the east.

the direction of the current, then the thumb will show the direction of the motion given to the conductor.

4. *To determine the direction of an induced current in a conductor that is moving in a magnetic field:* Place thumb, forefinger, and middle finger of the right hand at right angles to each other; if the forefinger shows the direction of the lines of force and the thumb shows the direction of motion of the conductor, then the middle finger will show the direction of the induced current.

A rule that is sometimes useful is the following: If the effect of the movement of a closed coil is to diminish the number of lines of force that pass through it, when viewed by a person looking along the magnetic field in the direction of the lines of force but if the effect is to increase the number of lines of force that pass through the coil—the current will flow in the opposite direction.

4

Power in AC and DC Circuits

Most circuits used in electrical systems for building construction are connected in parallel, and the total resistance in these circuits must sometimes be determined to calculate the current drawn or the power expended in them.

If both current and voltage are known, the easiest way to find the total resistance, in ohms, is by the use of Ohm's law, $R = E/I$. However, the resistance of a load is of little use to the practical worker except as a step toward finding more useful electrical units, such as current in amperes, voltage, or the wattage of a given load or circuit. Therefore, when problems dealing with resistance become apparent, usually only one other value in the circuit is known. This means that the total resistance of loads must be found by some other method.

The rule most often used to find the total resistance in a parallel circuit is the following:

The sum of the reciprocals of the separate resistors equals the reciprocal of the equivalent resistance.

Since the reciprocal of any number is 1 divided by the number, the equation for finding the total resistance of two or more resistors connected in parallel is

$$\frac{1}{R_t} = \frac{1}{R_1} + \frac{1}{R_2} + \frac{1}{R_3} + \text{etc.}$$

To illustrate, what is the total resistance of a parallel circuit whose resistors have a value of 30, 5, and 10 ohms, respectively? When adding fractions, they first must be changed to equal fractions, all having the same denominator. The fraction in the example may be changed so that all have 30 for the denominator:

$$\frac{1}{R_t} = \frac{1}{30} + \frac{6}{30} + \frac{3}{30} = \frac{10}{30}$$

The sum 10/30 is then the reciprocal of the resistance ($I/R_t = 10/30$). The reciprocal of any fraction is that fraction inverted. Then the reciprocal of 10/30 is 30/10, or 3. Therefore, the total resistance in the circuit is 3 ohms.

The total power in any dc circuit with only resistance loads may be found by the following equations: $P = E^2/R$, $P = I^2R$, and $P = EI$.

In an ac single-phase circuit containing motors, electric discharge lighting, relays, transformers, and the like, the equation is

$$P = E \times I \times \text{power factor}$$

For a practical application, assume that an ammeter is connected to an electrical circuit and shows a current reading of 17 amperes; a voltmeter is then connected and a reading of 240 volts is obtained. The load consists of electric heating units having pure resistance load. Using the equation $P = EI$, 240 volts × 17 amperes = 4080 watts.

If this same reading for both current and voltage were obtained on a motor circuit with an 85 percent power factor (PF), the wattage consumed would be 240 volts × 17 amperes × 0.85 PF = 3468 watts.

In a three-phase circuit, the voltage is always multiplied by the square root of 3, or 1.73, for use in circuit calculations. So if an electrical circuit draws 7.5 amperes of current per leg on a 480-volt circuit (between any two phase conductors), and the power factor is 80 percent, the total watts may be found by the equation

$$P = 1.73 \times E \times I \times \text{PF}$$

Substituting in the equation,

$$P = 1.73 \times 480 \times 7.5 \times 0.80 = 4982.4 \text{ watts}$$

RESISTANCE OF CONDUCTORS

It may be necessary during the course of an electrical installation to calculate the resistance of conductor lengths of a given size. A unit resistance of wire is the simplest way to calculate the resistance of wire length, and the unit *mil foot* is normally used. This unit represents a piece of round wire 1 mil in diameter and 1 foot in length; it is a unit that is accurate enough for all practical purposes. A round conductor is usually sized in circular mils (CM), which are determined by the diameter squared; that is, $d^2 = \text{CM}$.

The resistance of ordinary copper per circular mil foot will vary under various temperature conditions, but for all practical purposes 11 ohms per circular mil foot will suffice for copper wire and 18 ohms per circular mil foot for aluminum wire. These figures or *constants* are used when calculating voltage drop and should be remembered.

To determine the resistance of a length of No. 10 AWG copper wire, 50 feet in length, first multiply the length (in feet) of the wire by the constant 11 to obtain an answer of 550 ohms. Then refer to a table like the one in Table 4-1 to obtain the area of No. 10 AWG wire; it is 10,380 circular mils. Therefore, the first answer of 550 ohms must then be divided by 10,380 to obtain the answer of 0.0529 ohm, the resistance of 50 feet of No. 10 AWG copper wire.

The circular mil foot unit of 11 can also be used to calculate the resistance of a square or rectangular bus bar by using the conversion factor of 0.7854 to change the square mils to circular mils.

For instance, to obtain the resistance of a bus bar 1/4 × 2-1/2 inches, 100 feet in length, the following calculations would be used:

$$0.25 \text{ (thickness of bus bar in inches)} \times 1000$$
$$\text{(to convert inches to mils)} = 250$$

$$2.5 \text{ (width of bus bar)} \times 1000 = 2500$$

$$250 \times 2500 = 625,000 \text{ square mils}$$

Then to find the circular mil area, 625,000 square mils is divided by 0.7854 (conversion factor) to obtain an answer of 795,772.8 circular mils.

The calculation is continued in order to find the resistance of 100 feet of copper by multiplying 100 × 11 = 1100 and then dividing this by the circular mil area of the bus bar (795,778) = 0.0013 ohm.

SHORT-CIRCUIT CALCULATIONS

Several articles of the NE Code require that adequate interrupting capacity and protection be provided for all components on any electrical system. To provide this protection, the short-circuit currents at various intervals within the electrical distribution system must be calculated. The steps necessary for the calculation (assuming an unlimited primary short circuit) are as follows:

1. Determine the transformer full-load amperes from manufacturer's data, nameplate, or the following equations.
   (a) Three-phase transformer:

$$\text{Current} = \frac{\text{kVA} \times 1000}{\pm\text{line-to-line voltage} \times 1.73}$$

TABLE 4-1

Size AWG, MCM	Area Cir. Mils	Concentric Lay Stranded Conductors			Bare Conductors		DC Resistance Ohms/M Ft. At 25°C, 77°F.		
							Copper		Aluminum
		No. Wires	Diam. Each Wire Inches	Diam. Inches	Area Sq. Inches	Bare Cond.	Tin'd. Cond.	
18	1620	Solid	.0403	.0403	.0013	6.51	6.79	10.7
16	2580	Solid	.0508	.0508	.0020	4.10	4.26	6.72
14	4110	Solid	.0641	.0641	.0032	2.57	2.68	4.22
12	6530	Solid	.0808	.0808	.0051	1.62	1.68	2.66
10	10380	Solid	.1019	.1019	.0081	1.018	1.06	1.67
8	16510	Solid	.1285	.1285	.0130	.6404	.659	1.05
6	26240	7	.0612	.184	.027	.410	.427	.674
4	41740	7	.0772	.232	.042	.259	.269	.424
3	52620	7	.0867	.260	.053	.205	.213	.336
2	66360	7	.0974	.292	.067	.162	.169	.266
1	83690	19	.0664	.332	.087	.129	.134	.211
0	105600	19	.0745	.372	.109	.102	.106	.168
00	133100	19	.0837	.418	.137	.0811	.0843	.133
000	167800	19	.0940	.470	.173	.0642	.0668	.105
0000	211600	19	.1055	.528	.219	.0509	.0525	.0836
250	250000	37	.0822	.575	.260	.0431	.0449	.0708
300	300000	37	.0900	.630	.312	.0360	.0374	.0590
350	350000	37	.0973	.681	.364	.0308	.0320	.0505
400	400000	37	.1040	.728	.416	.0270	.0278	.0442
500	500000	37	.1162	.813	.519	.0216	.0222	.0354
600	600000	61	.0992	.893	.626	.0180	.0187	.0295
700	700000	61	.1071	.964	.730	.0154	.0159	.0253
750	750000	61	.1109	.998	.782	.0144	.0148	.0236
800	800000	61	.1145	1.030	.833	.0135	.0139	.0221
900	900000	61	.1215	1.090	.933	.0120	.0123	.0197
1000	1000000	61	.1280	1.150	1.039	.0108	.0111	.0177
1250	1250000	91	.1172	1.289	1.305	.00863	.00888	.0142
1500	1500000	91	.1284	1.410	1.561	.00719	.00740	.0118
1750	1750000	127	.1174	1.526	1.829	.00616	.00634	.0101
2000	2000000	127	.1255	1.630	2.087	.00539	.00555	.00885

Am. Gauge B. & S. No.	Diam. Mils	Circular Mils	Pounds Per 1000 Ft.	Pounds Per Mile	Feet Per Pound	Ohms Per 1000 Feet	Ohms Per Mile	Feet Per Ohm	Ohms Per Pound
0000	460.	211600.	639.8	3378.	1.56	.04906	.25903	20383.	.000076736
000	409.640	167805.	507.3	2678.	1.97	.06186	.32664	16165.	.00012039
00	364.800	133079.40	402.4	2124.	2.49	.07801	.41187	12820.	.00019423
0	324.950	105592.50	319.4	1686.	3.13	.09831	.51909	10409.	.00030772
1	289.300	83691.20	253.0	1336.	3.95	.12404	.65490	8062.3	.00048994
2	257.630	66373.	200.6	1059.	4.99	.15640	.82582	6393.7	.00078045
3	229.420	52634.	159.1	840.01	6.29	.19723	1.0414	5070.2	.0012406
4	204.310	41472.	126.2	666.3	7.93	.24869	1.3131	4021.	.0019721
5	181.840	33102.	100.0	528.2	10.	.31361	1.6558	3188.7	.0031361
6	162.020	26250.50	79.35	419.0	12.61	.39546	2.0881	2528.7	.0049868
8	128.490	16509.	49.92	263.6	20.05	.62881	3.3201	1590.3	.012608
9	114.430	13094.	39.57	208.9	25.28	.79281	4.1860	1261.3	.010042
10	101.890	10381.	31.30	165.8	31.38	1.	5.2800	1000.	.031380
12	80.808	6529.90	19.74	104.2	50.69	1.5898	8.3940	629.02	.080585
14	64.084	4106.76	12.42	65.59	80.42	2.5266	13.3405	375.79	.203180
16	50.820	2582.67	7.802	41.20	127.87	4.0176	21.2130	248.90	.513737
17	45.257	2048.20	6.20	32.746	161.24	5.0660	26.7485	197.39	.816839
18	40.303	1624.33	4.92	25.970	203.31	6.3380	33.7285	156.54	1.298764
19	35.890	1288.09	3.90	20.594	256.39	8.0555	42.5329	124.14	2.065312
20	31.961	1021.44	3.09	16.331	323.32	10.1584	53.6362	98.44	3.284374
21	28.462	810.09	2.45	12.952	407.67	12.8088	67.6302	78.07	5.221775
22	25.347	642.47	1.95	10.272	514.03	16.1504	85.2343	61.92	8.301819
23	22.571	509.45	1.54	8.1450	648.25	20.3674	107.540	49.10	13.20312
24	20.100	404.01	1.22	6.4593	817.43	25.6830	135.606	38.94	20.99405
25	17.900	320.41	.97	5.1227	1030.71	32.3833	170.984	30.88	33.37780
26	15.940	254.08	.77	4.0623	1299.77	40.8377	215.623	24.49	53.07946

No. B. & S.	Diameter		Square of Sectional Diameter = Area		Approved Carrying Capacity In Amperes National Code	Weight — Lbs. Bare		Approximate — Weatherproof Insulation	
	Mils	Mili-meters	Circular Mils	Square Millimeters		Lbs. Per 1000 Ft.	Lbs. Per Mile	Lbs. Per 1000 Ft.	Lbs. Per Mile
0000	460.	11.684	211600.	107.219	225	639.8	3378.	767	4050
000	409.640	10.405	167805.	85.028	175	507.3	2678.	629	3320
00	364.800	9.266	133079.4	67.431	150	402.4	2124.	502	2650
0	324.950	8.254	105592.5	53.504	125	319.4	1686.	407	2150
1	289.300	7.348	83694.2	42.409	100	253.0	1336.	316	1670
2	257.630	6.544	66373.	33.632	90	200.6	1059.	260	1370
3	229.420	5.827	52634.	26.670	80	159.1	840.1	199	1050
4	204.310	5.189	41742.	21.151	70	126.2	666.3	164	865
5	181.940	4.621	33102.	16.773	55	100.0	528.2	135	710
6	162.020	4.115	26250.5	13.301	50	79.35	419.0	112	590
8	128.490	3.264	16509.	8.366	35	49.92	263.6	75	395
10	101.890	2.588	10381.	5.260	25	31.30	165.8	53	280
12	80.808	2.053	6529.9	3.309	20	19.74	104.2	35	185
14	64.084	1.628	4106.8	2.081	15	12.42	65.59	25	130
16	50.820	1.291	2582.9	1.309	6	7.802	41.20	20	105

Figure 4-1   Characteristics of electrical conductors (continued on next page).

Size C. M.	7	19	37	7x7—49	61	91	127	169	217
			Diameter In Inches of Each Wire In The Strand						
2000000	.5345	.3243	.2325	.202	.181	.1482	.1255*	.1088	.096
1750000	.5000	.3034	.2175	.189	.1694	.1386	.1157*	.1003	.0898
1500000	.4629	.2810	.2013	.175	.1568	.1284*	.1087	.0942	.0831
1250000	.4226	.2565	.1838	.1507	.1431	.1172	.0992	.086	.0759
1000000	.378	.2294	.1644	.1429	.1285*	.1048	.0887	.0769	.0678
950000	.3684	.2236	.1602	.1392	.1248*	.1021	.0864	.075	.0662
900000	.3586	.2176	.1559	.1355	.1215*	.0994	.0841	.073	.0644
850000	.3484	.2115	.1516	.1317	.1181*	.0966	.0818	.0709	.0626
800000	.338	.205	.147	.1278	.1145*	.0937	.0793	.0687	.0607
750000	.3273	.1986	.1424	.1237	.1109*	.0908	.0768	.0666	.0588
700000	.3163	.1919	.1375	.1195	.1071*	.0883	.0742	.0644	.0568
650000	.3047	.1850	.1325	.1152	.1032*	.0845	.0716	.0620	.0547
600000	.2928	.1778	.1273	.1107	.0992*	.0812	.0687	.0596	.0526
550000	.2803	.1701	.1219	.106	.0950*	.0777	.0658	.0570	.0503
500000	.2672	.1622	.1162*	.101	.0905	.0741	.0627	.0544	.048
450000	.2535	.1539	.1103*	.0958	.0859	.0703	.0595	.0516	.0455
400000	.2391	.1451	.1040*	.0904	.081	.0663	.0561	.0487	.0429
350000	.2236	.1357	.0973*	.0845	.0757	.0620	.0526	.0455	.0401
300000	.207	.1257	.0901*	.0783	.0701	.0573	.0486	.0421	.0372
250000	.189	.1147	.0822*	.0714	.064	.0524	.0444	.0384	.0340
Size B. & S.									
0000	.1736	.1055*	.0756	.0657	.0589	.0482	.0408	....	....
000	.1548	.0940*	.0673	.0586	.0525	.0429	.0363	....	....
00	.1378	.0836*	.0599	.0521	.0467	.0382	.0323	....	....
0	.1228	.0746*	.0534	.0464	.0416	.0340	.0288	....	....
1	.1093	.0663*	.0475	.0413	.0370	.0303	.0252	....	....
2	.0973*	.0592	.0423	.0369	.0329	.0269	.0228	....	....
3	.0867*	.0526	.0377	.0327	.0294	.0240	.0203	....	....
4	.0772*	.0468	.0335	.0291	.0261	.0214	.0179	....	....
5	.0687*	.0417	.0299	.026	.0233	.0190	.0161	....	....
6	.0612*	.0372	.0266	.0231	.0207	.0169	.0143	....	....
7	.0545*	.0331	.0237	....	.0184	.0151	.0128	....	....
8	.0485*	.0293	.0211	.0184	.0164	.0135	.0114	....	....
9	.0435*	.0262	.0187	....	.0146	.0120	.0101	....	....
10	.0385*	.0223	.0168	....	.0129	.0106	.0090	....	....

*Indicates standard construction for rubber covered cables.

Total Capacity C. M.	2	3	4	5	6	Combinations	
1000000	....	....	....	0000	000	....	....
800000	....	....	0000	000	00	....	....
600000	....	0000	000	00	0	....	....
500000	....	000	00	0	1	....	....
400000	0000	00	0	1	2	....	....
350000	000	....	1	2	......	0000 & 00	....
300000	....	0	2	......	3	000 & 00	....
250000	00	1	....	3	4	000 & 1	0000 & 4
B. & S.							
0000	0	2	3	4	5	00 & 1	000 & 3
000	1	3	4	5	6	0 & 2	00 & 5
00	2	4	5	6	......	1 & 3	0 & 6
0	3	5	6	....	8	2 & 4	....
1	4	6	....	8	......	3 & 5	....
2	5	....	8	....	10	4 & 6	....
3	6	8	....	10	......	....	....
4	....	....	10	....	......	6 & 8	....
5	8	10	....	12	......	....	....
6	....	....	12	....	14	8 & 10	....
8	....	....	14	....	18	12 & 14	....
10	....	....	16	....	16	10 & 12	....

(b) Single-phase transformer:

$$\text{Current} = \frac{\text{kVA} \times 1000}{\text{line-to-line voltage}}$$

2. Find transformer multiplier by the equation

$$\text{Multiplier} = \frac{100}{\% \text{ transformer impedance}}$$

3. Determine transformer let-through short-circuit current from tables or by using the equation

$$I_{sca} = \text{transformer full-load amperes} \times \text{multiplier in step 2}$$

4. Determine f factor:
   (a) For three-phase faults:

$$f = \frac{1.73 \times \text{length of circuit to fault} \times \text{current}}{\text{constant} \times \text{line-to-line voltage}}$$

   (b) For single-phase, line-to-line faults on single-phase, center-tapped transformers:

$$f = \frac{2 \times \text{length of circuit to fault} \times \text{current}}{\text{constant (from table)} \times \text{line-to-line voltage}}$$

5. After solving for f, locate in the chart in Fig. 4-1 the proper multiplier M.
6. Multiply the available short-circuit current at the beginning of the circuit by multiplier M to determine the available symmetrical short-circuit current at the faults.

IMPEDANCE

Most of the electrical systems encountered in building construction for light and power contain inductance. In circuits with only pure resistance load, the inductance is usually small enough to be ignored. However, in circuits with motors, relays, transformers, electric discharge lighting, and the like, the inductance may be significant enough to be included in circuit calculations.

Current through an induction lags the voltage by 90°, and resistance (R) and the inductive reactance X_L, which provide opposition to the flow of current, may also be thought of as being 90° apart.

The total opposition to the flow of current is called impedance and may be represented by the hypotenuse of the triangle thus formed (see Fig. 4-2). Impedance, like other oppositions to the flow of current, is expressed in ohms. Impedance triangles may be solved with the principles of trigonometry. For example, if the impedance Z of the circuit illustrated in Fig. 4-3 is 5 ohms and angle A is $30°$, the inductive reactance X_L may be found as follows:

$$\sin A = \frac{\text{sides opposite}}{H}$$

$$\sin 30° = \frac{X_L}{5} = 2.5 \quad \text{and therefore} \quad X_L = 2.5$$

If X_L equals 2.5 ohms and the impedance Z is 5 ohms, the resistance may be found by

$$R = 5 \times \cos 30° \times 4.33 \text{ ohms}$$

When the resistance and inductive reactance in a circuit are known, the impedance may be found by the equation

$$Z = \sqrt{R^2 + X_L^2}$$

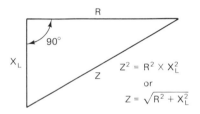

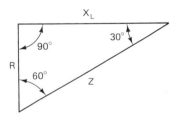

Figure 4-2 Impedance may be represented by the hypotenuse (Z) of a triangle.

Figure 4-3 Typical impedance triangle.

AC CIRCUIT CHARACTERISTICS

Impedance is the total opposition to the flow of alternating current. It is a function of resistance, capacitive reactance, and inductive reactance. The following equations relate these circuit properties:

$$X_L = 2\pi f C, \qquad X_C = \frac{1}{2\pi f C}, \qquad Z = \sqrt{R^2 + (X_L - X_C)}$$

where

X_L = inductive reactance in ohms
X_C = capacitive reactance in ohms
Z = impedance in ohms

f = frequency (cycles per second)
C = capacitance in farads
L = inductance in henrys
R = resistance in ohms
π = pi = 3.14

Ohm's law, when used in ac circuits, must include the total opposition to the flow of current. Therefore,

$$E = I \times Z, \qquad I = \frac{E}{Z}, \quad Z = \frac{E}{I}$$

POWER FACTOR

There are three different factors in all alternating-current circuits:

1. True power
2. Inductive reactance
3. Capacitive reactance

In any load or resistance in a circuit, a certain amount of power must be expended in the load of an inductive circuit. In inductive loads, current cannot be in phase with the voltage because of the reactance present. The amount of angular displacement between voltage and current depends on the ratio of resistance to reactance. Because the current and voltage do not reach maximum at the same instant, the real power of the circuit must be $E \times I$ multiplied by some factor less than 1. This factor compensates for the phase displacement between voltage and current. This factor is called the *power factor of the load* and is equal to the cosine of the angle between the voltage across and the current drawn by the load. The power factor is also equal to the ratio of load resistance to load impedance. The equation, in simplified form, for determining the power factor is

$$PF = \frac{\text{kilowatts}}{\text{kilovolt-amperes}}$$

An analogy will enhance the understanding of power factor. Imagine a farm wagon on a country road to which three horses are hitched as shown in Fig. 4-4. The one in the middle is pulling straight down the road. We call him "True Power" because all his effort is in the direction that work should be done. The one on the right, horse 2, is a fickle small animal. He always wants to nibble at the daisies on his side of the road. He does not contribute an ounce of pull in the desired direction but causes havoc by pulling us toward the ditch. We call him Mr. Inductance. Horse 3, the size of horse 2,

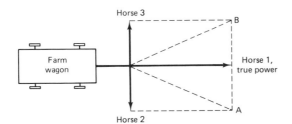

Figure 4-4   An analogy used to give an understanding of power factor.

is 180° out of phase with horse 2 and seems to enjoy the succulent grass on the left side. He is as opposite from horse 2 as north is from south. He also has a right-angle complex and contributes nothing to forward motion. We call him Mr. Capacitance. Incidentally, we would never have horses 2 or 3 hitched to our wagon if we were analyzing a dc circuit. They just cannot exist in a dc circuit.

For the moment we will forget horse 3. If only horses 1 and 2 were pulling, the wagon would go in the direction of the dashed line A. Notice the length of that line is greater than the line to horse 1, so Mr. Inductance has an effect on the final result. The direction and length of A might well be called *apparent power,* and it is the hypotenuse (or diagonal) of a right-angle triangle.

If horse 2 were left at home, then horses 1 and 3 in combination would pull toward B and the length of that line would also be apparent power. If all three were on the job, horses 2 and 3 would cancel one another, and the only useful animal is reliable horse 1.

In an ac circuit we always have all three forces in the picture in varying length. We should not complain about inductance for it is always present in every magnetic circuit and always works in a 90° angle with True Power. Without magnetism, we would have no motors or transformers.

CAPACITORS

A capacitor produces neither heat or magnetism, but it does have several useful characteristics in an electrical circuit. The capacitive current will lead the voltage by 90 electrical degrees, whereas a pure inductor lags its voltage by 90°.

Capacitors have the property to oppose any change in voltage. They also have the ability to store electrical energy and are used extensively in all types of electrical and electronic circuits.

When connected to a dc power supply, current cannot pass through, but because of the large area of conducting material, the two surfaces charge up to the supply voltage in a short time interval. A capacitor can hold this charge for an indefinite period of time after being disconnected from the supply voltage. It will discharge through an external circuit if one is provided. The capacitor discharge takes place very quickly.

Capacitors limit or oppose ac because the voltage magnitude is constantly changing. The magnitude of this opposition is dependent on the physical size of the capacitor, the insulating material or dielectric, and how much area of the conducting materials it contains. An evaluation of a capacitor's opposition in ohms may be made by the equation

$$X_C = \frac{1}{2\pi f C}$$

where X is its capacitive reactance measured in ohms, C is its capacitance in farads, and f is the frequency in hertz of the ac applied to the capacitor.

A farad is a very large unit; therefore, capacitance is usually measured in microfarads. Then the equation may be written

$$X_C = \frac{10^6}{2\pi f C} = \frac{159 \times 10^3}{f C}$$

where C is capacitance measured in microfarads (μF), or millionths of a farad.

The current through a capacitor can be calculated in the same manner as for the resistor or the inductor by dividing the voltage across the capacitor by its capacitive reactance in ohms, or

5

Single-Phase Systems

A single-phase, alternating-current system has a single voltage in which voltage reversals occur at the same time and are of the same alternating polarity throughout the system as shown in Fig. 5-1.

PRODUCTION OF SINGLE-PHASE ALTERNATING CURRENT

Referring to Figs. 5-2, 5-3, and 5-4 will help you to understand how single-phase, alternating current is produced. The blocks on each side of the moving coil *(M)* represent powerful magnets between which invisible magnetic lines of force (flux) pass horizontally. The wire loop, rectangualr in form, rotates between these two magnets, and each end of the loop is connected to slip rings *A* and *B*. Assume that the wire loop is rotating on its own axis 3600 times per minute or 60 revolutions per second in a clockwise direction as viewed from the slip rings. If the flux lines passed from the left pole to the right pole (*M* to *M* from left to right), then at the instant the wire is in the position shown the current in it would flow from front to back. Also at this precise instant of time, the wire is cutting the maximum number of magnetic lines of force, thus generating the highest voltage and current. Simultaneously, the wire partially concealed by the right pole is moving through the same flux field but in the opposite direction. This causes the generated current to flow from the back to the front. Thus, the currents are in the same direction in the loop and assist one another. At this instant in time, slip ring *A* is negative and *B* is positive.

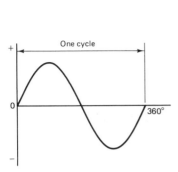

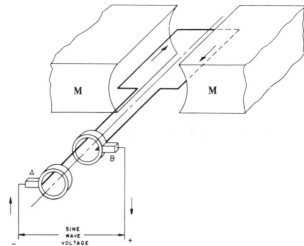

Figure 5-1   One cycle of a single-phase alternating current.

Figure 5-2   Simple electric generator consisting of a wire loop revolving in a strong magnetic field.

At the instant described in the preceding paragraph, the top and bottom conductors are passing between lines of flux. Therefore, no voltage or current is being generated. In Fig. 5-4, another revolution has elapsed, so apparently we are under the same condition as in Fig. 5-2. However, there has been a change this time! Even though the current direction is as before, note carefully that the slip ring *B* is now connected to the left side of the moving coil and has therefore changed from positive to negative.

At any point between 1 and 2 (Fig. 5-5), the magnitude of the generated current (or voltage) is becoming smaller and smaller as fewer and fewer lines

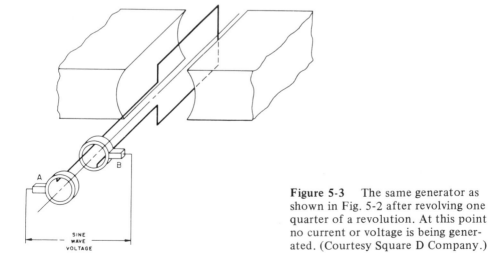

Figure 5-3   The same generator as shown in Fig. 5-2 after revolving one quarter of a revolution. At this point no current or voltage is being generated. (Courtesy Square D Company.)

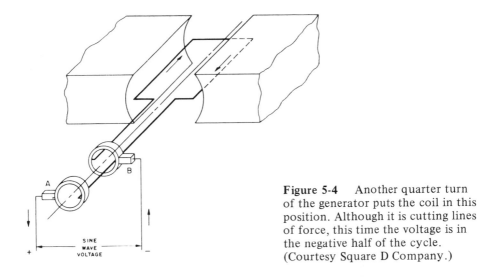

Figure 5-4 Another quarter turn of the generator puts the coil in this position. Although it is cutting lines of force, this time the voltage is in the negative half of the cycle. (Courtesy Square D Company.)

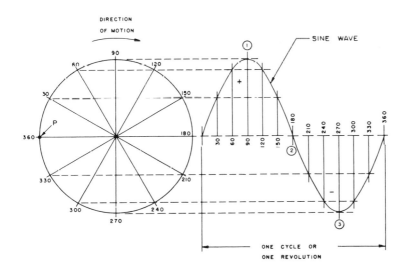

Figure 5-5 Net result of a voltage or current wave created by a generator. (Courtesy Square D Company.)

of flux are cut. Finally, zero current is reached at 2 as shown in Fig. 5-5. Between 2 and 3, the current builds up and this phenomenon repeats itself 120 times every second. The net result is voltage, or a current wave, which creates a picture as in Fig. 5-5.

In the verbal explanation of zero or maximum currents and voltages, we selected only two positions of the flat coil. The circled numbers 1, 2, and 3 represent the voltage generated in Figs. 5-2, 5-3, and 5-4, respectively. Obviously, there are theoretically an infinite number of positions in any revolution, but for our purpose we have selected positions every 30° around the full circle of rotation. When these points are plotted, they create a smooth curve that is known as a *sine wave,* because the shape of this curve is identical to another curve drawn from the sine functions in trigonometry.

The maximum value or height of the voltage wave is dependent on the constant speed of generator rotation and the strength of the magnetic field. For example, if the distribution system was designed for 2400 volts between the high-voltage wires out on the line, the plant operator could increase the magnetic field current enough to boost the voltage on the plant bus to 2450 volts. This would assure that out at the end of a long distribution line the last customer's transformer would receive a voltage adequate to provide the house voltage within the limits set by the Public Service Commission.

The wave in question is called a 60-cycle or 60-hertz wave (the latter term is used more frequently these days), because when the generator turns at its rated speed, it will produce exactly 60 identical curves that repeat themselves each second. Hence a cycle is one complete wave, which, when it begins to repeat itself, is the start of a new cycle.

Around the turn of the century, or perhaps a little before, many generating plants were operating at 25 cycles, that is, a 25-cycle system. However, incandescent lamps tried to go out each time the current passed through zero, causing the lamps to flicker, which was very annoying. At 60 cycles no flicker is discernible.

Many foreign countries still utilize a 25-hertz system. Canada used such a system up to about the early 1950s, but has since changed over to 60 hertz exclusively. Obviously, a 25-hertz system cannot be interchanged with a 60-hertz system.

SINGLE-PHASE CONNECTIONS

In the early days of residential electrical service, it was quite common to supply only 30-ampere, 120-volt single-phase service, as lighting was the primary load with the occasional use of small 120-volt appliances such as a radio, toaster, iron, and the like. For this type of service, a transformer was placed between the high-voltage line and the line containing low voltage, as shown in Fig. 5-6. The 120/240-volt low-voltage windings are connected in parallel, giving 120 volts on a two-wire system. The connection, shown in

Fig. 5-7, is similar to the one shown previously except that the windings are connected in series to produce 240 volts on a two-wire system.

The connection shown in Fig. 5-8 is the most common single-phase distribution system in use today. It is known as the three-wire, 240/120-volt single-phase system and is used where 120 and 240 volts are used simultaneously.

Two single-phase circuits operating 90° out of phase with each other are called a two-phase system. Two different transformer connections are normally used. The one shown in Fig. 5-9 is treated as two separate circuits, while the connection in Fig. 5-10 has a common wire on the secondary side, resulting in some saving of copper wire. In the latter connection, the common wire must carry 1.41 times the current in other secondary wires.

When an alternator is constructed to generate single-phase voltage, all stator coils are connected in series to form one closed circuit. However, most alternators are wound for three-phase operation. Therefore, the stator coils are connected to form three distinct and independent windings that are displaced 120° from each other. The total voltage generated in the coils of one such winding, or phase, is called the *phase voltage* of the alternator. The three windings are connected either to a common point to build a star-connected alternator or end to end to form a delta-connected alternator. As a rule, three terminal leads are brought out from the interconnected windings. In a star-connected alternator, the neutral point can also be brought out to be grounded for protection. According to ac circuit theory, the delta-connected alternator has line voltage equal to the phase voltage, and in the star-connected alternator the line voltage is 1.732 times the phase voltage.

The rotor commonly carries the poles, or the field. The number of poles and the form of the rotor depend on the speed at which the rotor will revolve. The two forms of rotors are the salient pole and the round, or cylindrical, rotor.

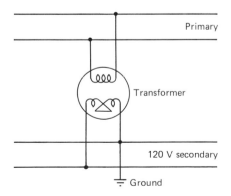

Figure 5-6 Transformer connection giving single-phase, 120-volt service. (Courtesy Square D Company.)

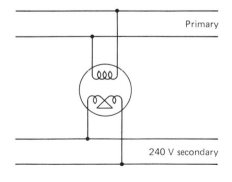

Figure 5-7 Transformer connection giving single phase, 240 volts on a two-wire system.

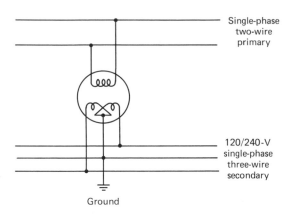

Figure 5-8 Transformer connection giving 120/240 volts, single phase on a three-wire system.

A salient-pole rotor carries several laminated pole cores that are bolted or dovetailed to a laminated rotor body. All rotor coils are wound in the same direction. In order that adjacent poles will be of opposite polarity, current flows from top to bottom in one coil and from bottom to top in the adjacent coil. The leads that connect the coils go from the bottom of one coil to the bottom of the next and from the top of that coil to the top of the coil that follows.

In some alternators there is a pole-face winding in addition to the field winding. Pole-face windings, called damper, or amortisseur, windings, consist of copper or bronze bars inserted in holes provided for them in the punching near the air-gap surface of the pole. The bars in any one pole are connected together at their ends by brazing them to the pole end plates or to a short-circuiting bar. Occasionally they are connected from pole to pole. The bars are not insulated from the poles. Damper windings serve two purposes: to dampen the fluctuations in speed and to prevent excessive unbalanced voltages during short circuits. The damper winding is important when the prime mover is a diesel or other reciprocating type of engine.

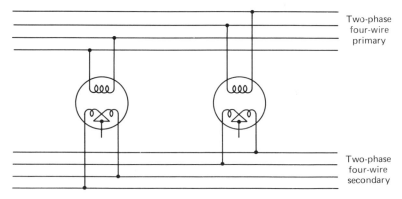

Figure 5-9 Two-phase, four-wire secondary system.

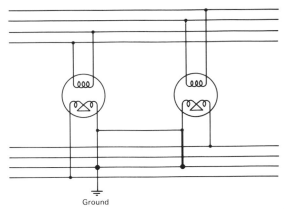

Ground

Figure 5-10   Common wire used
on a two-phase system.

Alternators that use steam turbines as prime movers run at high speeds and, instead of salient-pole rotors, use cylindrical or round rotors. A round rotor is made comparatively narrow and long in order to withstand the mechanical stresses that are due to the high speed. It also has the field windings placed in radial slots in the rotor body, instead of being wound around salient-pole bodies. Several coils connected in series are used to provide the necessary magnetic force.

Hard copper strap bent on edge is used for coils, although some designs call for aluminum alloy. The conductors are bare, and turns are insulated from each other by strips of mica or a similar insulating material.

The ends of the field winding are connected to slip rings, or collector rings, mounted on and insulated from the shaft. Stationary carbon brushes held against the surface of the rings conduct the direct field current from the exciter through the rotating rings to the field coils.

Brushes are made of a suitable material, usually a relatively soft electrolytic graphite. They are supported in brush holders that keep them at the proper angle to the collector rings. The brushes are free to move radially in the holders, while spring pressure keeps them firmly on the rings. Proper brush pressure is necessary for satisfactory brush performance. A brush spring must be capable of accurate adjustment of brush pressure, and it must maintain constant pressure over the full wearing length of the brush. The brush holder and the spring assembly are usually supported on an insulated stud from the housing. The number of brushes depends on the amount of exciting current the brushes are to carry. Attached to each brush is a flexible braided-wire pigtail, or shunt, which is connected to the brush holder or to a copper bus bar to provide a low-resistance path for current from the brush to the external circuit.

6

Three-Phase Systems

A three-phase, alternating-current system has three individual circuits or phases. Each phase is timed so that the current alternations of the first phase are one-third cycle (120°) ahead of the second and two-thirds cycle (240°) ahead of the third, as shown in Fig. 6-1.

All large blocks of power at high-voltage transmission and at distribution voltage are transmitted three phase. The material presented in Chapter 5 on single-phase systems will be helpful in understanding how three-phase systems operate, because, in the simplest concept, three-phase power is three single-phase power sources working together in an orderly and efficient combination.

OPERATION OF A THREE-PHASE SYSTEM

In a single-phase system, a single generator armature coil creates a complete cycle of current and voltage for each revolution. Picture three such loops going to separate slip rings so that loads could be connected to them. In reality, three separate single-phase generators wound on a single rotating armature would be the result.

If three separate coil conductors are equally spaced around the generator armature, their voltage waves will be separated by 120 electrical degrees. Referring again to Fig. 6-1, the first wave is called phase A; the next wave phase B would start its rise at 120°, and phase C starts its rise at 240°.

Three-phase power may also be described by using a vector diagram, such as the one in Fig. 6-2. In this diagram, NA is the voltage vector for phase A, and NB and NC are the corresponding vectors for phases B and C.

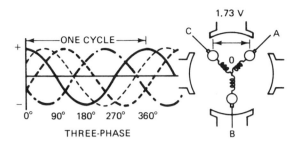

THREE-PHASE

Figure 6-1

They are spaced 120° in a conventional counterclockwise rotation. By electrically connecting the other ends of these coils at a common point, N (common neutral), the diagram as shown is created. Because this diagram looks like the letter Y, this type of three-phase connection is known as the *wye* connection. Sometimes it is referred to as the *star* connection.

The diagram in Fig. 6-2 creates other valuable information. If the generator produced 120 volts from N to A, it also would produce the same voltage from N to B and from N to C. If these lengths were drawn to scale and were measured, they would equal 120 volts. However, if a straight line were drawn between A and B, B and C, and C and A, all would scale 208 volts, which is actually the voltage between these phase conductors. So on a wye-connected three-phase system, we can see that the voltage between the neutral (N) and any one phase conductor (A, B, or C) will be 120 volts, if this is the voltage produced by the generator. However, when a reading is taken across phases (A to B, B to C, or C to A), the voltage will be 208 volts. Thus, a three-phase, 120/208-volt wye system is obtained.

Deriving these values by drawing pictures or diagrams is tedious and time consuming. An easier way is to merely multiply the phase-to-neutral voltage by the square root of 3, which is 1.732. Thus, if NA was 277 volts, then AB = 480 volts. Similarly, a phase-to-neutral voltage of 2400 creates 4160 phase to phase, and 7200 phase to neutral becomes 12,470 volts. These voltage combinations become second nature to persons who have been involved in the trade for some time.

One of the benefits that a wye distribution system brings to the utility or municipality is that even though their turbogenerators are rated at 2400 and 7200 volts, they can transmit at a 72 percent higher phase-to-phase voltage with reduction in losses and better voltage regulation at the ends of longer rural lines. Most distribution in the United States is at 12,470/7200 volts to gain this benefit. In these systems, the neutral junction point is grounded, and a fourth wire is carried along the system and grounded at every distribution transformer location. This solidly grounded wye is regarded as the safest of all systems and also provides the best assurance of proper operation of the protective devices needed to isolate sections of a system in case of trouble.

Delta Systems

The first three-phase distribution system was the delta system, for which a vector diagram is shown in Fig. 6-3. In this system, the generator windings are connected as shown, and the junction points are identified as phases *A, B,* and *C.* Only three wires appear on the distribution system. Because none of the three lines are usually connected to ground, an artificial ground midway between the three points of the triangle tends to exist. It is free to shift when the loads on each phase are not equal. If a measurement was made to ground, the value would be the phase-to-phase voltage divided by the square root of 3. Therefore, for 7200 volts, phase to phase, the reading to ground would be 4160 volts.

In either wye or delta systems, three-phase transformers are connected to phases *A, B,* and *C.* Single-phase loads on wye systems are usually connected from phase to neutral, or they could be fed between the phase wires if high-voltage transformers are available. In the delta system, single-phase loads are always connected between the phase wires.

One connection that is possible in a three-phase delta system is known as *open delta.* This connection permits the availability of three-phase power anywhere along the distribution line with the use of only two transformers rather than with the usual three units. This connection reduces capital investment and adds only one penalty: the three-phase load that can be carried by an open-delta bank is only 86.6 percent of the combined rating of the two equal-sized units. It is only 57.7 percent of the nominal full-load capability of a full bank of transformers. In an emergency, however, this capability permits single- and three-phase power at a location where one unit has burned out and a replacement is not readily available. In such a situation, the total load must be curtailed to avoid additional burnouts. Therefore, such a connection is never expected to carry the nameplate ratings for the transformers.

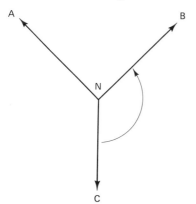

Figure 6-2   Vector diagram of wye three-phase system.

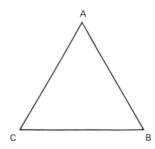

Figure 6-3   Vector diagram of delta distribution system.

When delta banks of transformers are not grounded, it is possible for one phase to accidentally become grounded without the operators being aware of this. Not until another phase also grounds out will the problem be apparent. A few power companies deliberately ground one corner of their delta system so that inadvertent faults on the other two phases will cause fuses, reclosers, or circuit breakers to clear the fault.

Some power companies ground the midway point between the two phases on delta systems, permitting some interesting voltage possibilities, as shown in Fig. 6-4. Three-phase power is available as normal between phases *A, B,* and *C.* Single-phase 240 volts can be had between any pair of phase conductors, while single-phase 120-volt power may be connected between ground and either *B* or *C* phase. Finally, 208 volts, single phase, is possible between ground and *A* phase.

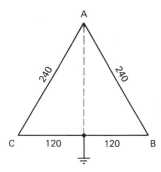

Figure 6-4   Delta system with ground connection halfway on one phase.

Figure 6-5 shows some of the more common ways in which three-phase transformer windings can be connected. In general, they are as follows:

Primary Winding	Secondary Winding
Delta	Delta
Delta	Wye or star
Wye	Delta
Wye	Wye

The connections in Fig. 6-5 should be self-explanatory, and all of them except the last two should give little problems. The last two, however, the ones with the wye primary, may present some hazards. Here's why.

In a three-phase transformer, all three primary coils produce flux, and the flux from each coil is distributed through all three of the core legs. The secondary coils under load also produce flux, distributed through the same core legs in the same manner, and this flux is in opposition to the primary flux, tending to cancel it out. In a properly designed transformer, essentially

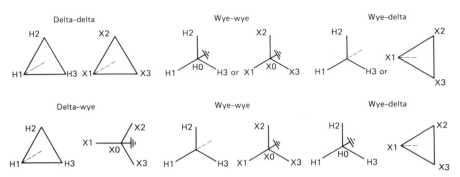

Figure 6-5   Three-phase transformer connections. (Courtesy Square D Company.)

all the flux travels through the core in just the right amount to produce the needed magnetic induction.

Now take a three-phase transformer with a wye-connected primary and a grounded neutral. The secondary can be connected either delta or wye—it makes no difference. Suppose for some reason, such as a wind storm, a wire that was feeding one of the primary coils became disconnected. The grounded neutral will allow three-phase flux to be produced, even though one of the primary coils is open. Obviously, the open primary produces no flux, but its secondary is still connected to the load and will be carrying a current. This secondary current produces a secondary flux, usually an excessive amount; and since there is no opposing flux, this flux will quickly fill up the core (it is called *saturation*). Any excess will take the path of least resistance to complete the magnetic circuit. Because large transformers are made with structural steel bases, angles, and enclosure panels, they become an ideal path for the leakage flux. They can become very hot, enough to blister the paint or even cause fire, not to mention the obvious damage to a very expensive transformer. For this reason, wye primaries with grounded neutrals are definitely not recommended unless some means is provided to disconnect all three phases in case of a fault or open circuit in the primary circuit.

7

Transformers

The electric power produced by alternators in a generating station is transmitted to locations where it is utilized and distributed to users. Many different types of transformers play an important role in the distribution of electricity. Power transformers are located at generating stations to step up the voltage for more economical transmission. Substations with additional power transformers and distribution transformers are installed along the transmission line. Finally, distribution transformers are used to step down the voltage to a level suitable for utilization.

TRANSFORMER BASICS

A very basic transformer consists of two coils or windings formed on a single magnetic core. Such an arrangement will allow transforming a large alternating current at low voltage into a small alternating current at high voltage, or vice versa. Transformers, therefore, enable changing or converting power from one voltage to another. For example, generators that produce moderately large alternating currents at moderately high voltages utilize transformers to convert the power to very high voltage and proportionately small current in transmission lines, permitting the use of smaller cable and providing less power loss.

The elementary principle of a transformer is shown in Fig. 7-1. Two windings are shown on a rectangular core made of iron. The source of alternating current and voltage is connected to the primary winding of the transformer. The secondary winding is connected to the circuit in which

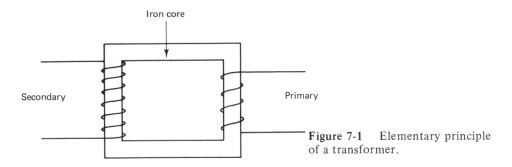

Figure 7-1   Elementary principle of a transformer.

there is to be a higher voltage and smaller current, although it could be a higher voltage and lower voltage. If there are more turns on the secondary than on the primary winding, the secondary voltage will be higher than that in the primary and by the same proportion as the number of turns in the winding. The secondary current, in turn, will be proportionately smaller than the primary current. With fewer turns on the secondary than on the primary, the secondary voltage will be proportionately lower than that in the primary, and the secondary current will be that much larger. Since alternating current continually increases and decreases in value, every change in the primary winding of the transformer produces a similar change of flux in the core. Every change of flux in the core and every corresponding movement of magnetic field around the core produce a similarly changing voltage in the secondary winding, causing an alternating current to flow in the circuit that is connected to the secondary.

Theoretical study of conditions in a transformer includes the use of phasor diagrams that represent graphically voltages and currents in transformer windings. Calculations of impedance can be simplified by using equivalent circuits. Reactance voltage drop is governed by the leakage flux, and the voltage regulation depends on the power factor of the load. To determine transformer efficiency at various loads, it is necessary to first calculate the core loss, hysteresis loss, eddy-current loss, and load loss.

Distribution transformers are used in transmission and distribution systems to provide the desired quantities of voltage. Most are rated up to 500 kilovolt-amperes, have steel cores, and use cylindrical coils for voltages up to about 25,000 volts; disk coils are used for higher voltages. Distribution transformers are protected by primary fuses, lightning arresters, surge diverters, and proper grounding. They are available as residential pole-mounted or pad-mounted transformers, rural-line transformers, and unit substations that contain all auxiliary equipment and may be stationary or portable.

Power transformers are designed for higher voltages and kilovolt-ampere ratings than distribution transformers. Their cores are constructed either in steel or core form, and the higher-voltage windings generally consist of disk coils. The shielding of coils helps to improve voltage distribution under voltage surges, while temperature indicators and thermal relays give a warning or disconnect the transformer when the temperature rises too high within the transformer. Almost all power transformers are oil filled and can be self-cooled, forced-air cooled, water cooled, and forced-oil cooled, depending on specific conditions. Transformer oil needs careful attention because of expansion, breathing action, deterioration due to moisture, and fire and explosion hazards. Inert-gas systems and activated alumina filters protect the oil.

There are transformers with only one winding, known as air-core or iron-core reactors. Constant-current transformers are designed to provide constant current regardless of load variations. Some transformers are designed for specific applications, such as control transformers and test transformers.

TRANSFORMER CHARACTERISTICS

In a well-designed transformer, there is very little magnetic leakage. The effect of the leakage is to cause a decrease of secondary voltage when the transformer is loaded. When a current flows through the secondary in phase with the secondary voltage, a corresponding current flows through the primary in addition to the magnetizing current. The magnetizing effects of the two currents are equal and opposite.

In a perfect transformer, that is, one having no eddy-current losses, no resistance in its windings, and no magnetic leakage, the magnetizing effects of the primary load current and the secondary current neutralize each other, leaving only the constant primary magnetizing current effective in setting up the constant flux. If supplied with a constant primary pressure, such a transformer would maintain constant secondary pressure at all loads. Obviously, the perfect transformer has yet to be built; the closest is one with very small eddy-current loss where the drop in pressure in the secondary windings is not more than 1 to 3 percent, depending on the size of the transformer.

TRANSFORMER TAPS

If the exact rated voltage could be delivered at every transformer location, transformer taps would be unnecessary. However, this is not possible, so taps are provided on the secondary windings to provide a means of either increasing or decreasing the secondary voltage.

Generally, if a load is very close to a substation or power plant, the voltage will consistently be above normal. Near the end of the line the voltage may be below normal.

In large transformers, it would naturally be very inconvenient to move the thick, well-insulated primary leads to different tap positions when changes

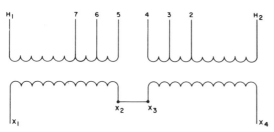

Figure 7-2   Typical use of taps, shown here in schematic form.

in source-voltage levels make this necessary. Therefore, taps are used, such as shown in the wiring diagram in Fig. 7-2. In this transformer, the permanent high-voltage leads would be connected to H_1 and H_2, and the secondary leads, in their normal fashion, to X_1 and X_2, X_3, and X_4. Note, however, the tap arrangements available at taps 2 through 7. Until a pair of these taps is interconnected with a jumper wire, the primary circuit is not completed. If this were, say, a typical 7200-volt primary, the transformer would have a normal 1620 turns. Assume 810 of these turns are between H_1 and H_6 and another 810 between H_3 and H_2. Then, if taps 6 and 3 were connected together with a flexible jumper on which lugs have already been installed, the primary circuit is completed, and we have a normal ratio transformer that could deliver 120/240 volts from the secondary.

Between taps 6 and either 5 or 7, 40 turns of wire exist. Similarly, between taps 3 and either 2 or 4, 40 turns are present. Changing the jumper from 3 to 6 to 3 to 7 removes 40 turns from the left half of the primary. The same condition would apply on the right half of the winding if the jumper were between taps 6 and 2. Either connection would boost secondary voltage by 2-1/2 percent. Had taps 2 and 7 been connected, 80 turns would have been omitted and a 5 percent boost would result. Placing the jumper between taps 6 and 4 or 3 and 5 would reduce the output voltage by 5 percent.

TRANSFORMER POLARITY

The polarity of a transformer is an indication of the direction of flow of current from a terminal at any one instant. The idea is quite similar to the polarity marking on a battery.

The polarity of a transformer is determined by the lead markings (see Fig. 7-3). When X_1 is located diagonally with respect to H_1, the polarity is, by definition, additive. When H_1 and X_1 are adjacent, the polarity is said to be subtractive.

As you face the high-voltage side of a transformer, the high-voltage terminal on your right is always marked H_1 and the other high-voltage terminal is marked H_2. This is an established standard.

In making transformer connections, particularly bank connections, the polarity of individual transformers must be checked. In making such connections, it is necessary to remember that all H_1 terminals are of the same polarity and all X_1 terminals are of the same polarity. So if you were to connect two single-phase transformers in parallel, you should connect the two H_1 terminals together and the two X_2 transformer terminals together. By following this procedure, two transformers can be satisfactorily paralleled regardless of whether they are both of the same polarity or whether one is additive and one is subtractive polarity.

The ANSI standards for transformers specify additive polarity as standard on all single-phase units in sizes of 200 kilovolt-amperes and smaller having high-voltage windings of 8600 volts and below. All other units are subtractive polarity.

TRANSFORMER CONNECTIONS

Transformer connections are many, and space does not permit the description of all of them here. However, an understanding of a few will give the basic requirements and make it possible to use manufacturer's data for others should the need arise.

Single-Phase Light and Power

The diagram in Fig. 7-4 is a transformer connection used quite extensively for residential and small commercial applications. It is the most common single-phase distribution system in use today. It is known as the three wire,

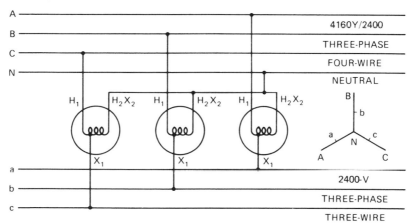

Figure 7-3   When X1 is located diagonally with respect to H1, the polarity is said to be additive. When H1 and X1 are adjacent, the polarity is said to be subtractive.

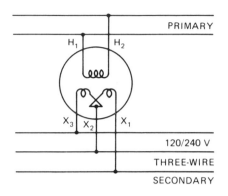

PRIMARY

H_1 H_2

X_3 X_2 X_1

120/240 V

THREE-WIRE

SECONDARY

Figure 7-4   Transformer connection
used quite extensively for residential
and small commercial applications.

240/120-volt single-phase system and is used where 120 and 240 volts are
used simultaneously.

Y-Y for Light and Power

The primaries of the transformer connection in Fig. 7-5 are connected in Y.
When the primary system is 2400/4160Y volts, a 4160-volt transformer is
required when the system is connected in delta-Y. However, with a Y-Y system,
a 2400-volt transformer can be used, offering a saving in transformer cost. It
is necessary that a primary neutral be available when this connection is used,
and the neutrals of the primary system and the transformer bank are tied
together as shown in the diagram. If the three-phase load is unbalanced, part
of the load current flows in the primary neutral. For these reasons, it is
essential that the neutrals be tied together as shown. If this tie were omitted,
the line to neutral voltages on the secondary would be very unstable. That is,
if the load on one phase were heavier than on the other two, the voltage on
this phase would drop excessively and the voltage on the other two phases
would rise. Also, large third-harmonic voltages would appear between lines
and neutral, both in the transformers and in the secondary system, in addition
to the 60-hertz component of voltage. This means that for a given value of
rms voltage, the peak voltage would be much higher than for a pure 60-hertz
voltage. This overstresses the insulation both in the transformers and in all
apparatus connected to the secondaries.

PARALLEL OPERATION OF TRANSFORMERS

Transformers will operate satisfactorily in parallel on a single-phase, three-
wire system if the terminals with the same relative polarity are connected
together. However, the practice is not very economical because the individual
cost and losses of the smaller transformers are greater than one larger unit

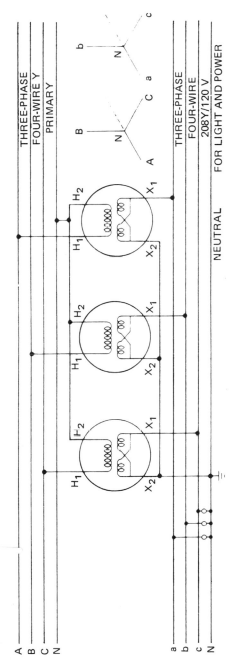

Figure 7-5 Y-Y three-phase transformer connection for light and power.

giving the same output. Therefore, paralleling of smaller transformers is usually done only in an emergency. In large transformers, however, it is often practical to operate units in parallel as a regular practice.

Two three-phase transformers may also be connected in parallel provided they have the same winding arrangement, are connected with the same polarity, and have the same phase rotation.

AUTOTRANSFORMERS

An autotransformer is a transformer whose primary and secondary circuits have part of a winding in common and therefore the two circuits are not isolated from each other. The application of an autotransformer is a good choice for some users where a 480Y/277- or 208Y/120-volt, three-phase, four-wire distribution system is utilized. Some of the advantages are as follows:

1. Lower purchase price
2. Lower operation cost due to lower losses
3. Smaller size; easier to install
4. Better voltage regulation
5. Lower sound levels

An autotransformer, however, cannot be used on a 480- or 240-volt, three-phase, three-wire delta system. A grounded neutral phase conductor must be available in accordance with NE Code Article 210-9, which states:

> *210-9.* Circuits Derived from Autotransformers. Branch circuits shall not be supplied by autotransformers.
>
> *Exception No. 1:* Where the system supplied has a grounded conductor that is electrically connected to a grounded conductor of the system supplying the autotransformer.

CONTROL TRANSFORMERS

Control transformers are available in numerous types, but most control transformers are dry-type step-down units with the secondary control circuit isolated from the primary line circuit to assure maximum safety. These transformers and other components are usually mounted within an enclosed control box or control panel, which has a push-button station or stations independently grounded as recommended by the NE Code. Industrial control transformers are especially designed to accommodate the momentary current inrush caused when electromagnetic components are energized, without sacrificing secondary voltage stability beyond practical limits.

Other types of control transformers, sometimes referred to as control and signal transformers, normally do not have the required industrial control transformer regulation characteristics. Rather, they are constant-potential, self-air-cooled transformers used for the purpose of supplying the proper reduced voltage for control circuits of electrically operated switches or other equipment and, of course, for signal circuits. Some are of the open type with no protective casing over the winding, while others are enclosed with a metal casing over the winding.

In selecting control transformers for any application, the loads must be calculated and completely analyzed before the proper transformer selection can be made. This analysis involves every electrically energized component in the control circuit. To select an appropriate control transformer, first determine the voltage and frequency of the supply circuit. Then determine the total inrush volt-amperes (watts) of the control circuit. In doing so, do not neglect the current requirements of indicating lights and timing devices that do not have inrush volt-amperes, but are energized at the same time as the other components in the circuit. Their total volt-amperes should be added to the total inrush volt-amperes.

POTENTIAL TRANSFORMERS

In general, a potential transformer is used to supply voltage to instruments such as voltmeters, frequency meters, power-factor meters, and watt-hour meters. The voltage is proportional to the primary voltage, but it is small enough to be safe for the test instrument. The secondary of a potential transformer may be designed for several different voltages, but most are designed for 120 volts. The potential transformer is primarily a distribution transformer especially designed for good voltage regulation so that the secondary voltage under all conditions will be as nearly as possible a definite percentage of the primary voltage.

CURRENT TRANSFORMER

A current transformer is used to supply current to an instrument connected to its secondary, the current being proportional to the primary current, but small enough to be safe for the instrument. The secondary of a current transformer is usually designed for a rated current of 5 amperes.

A current transformer operates in the same way as any other transformer in that the same relation exists between the primary and the secondary current and voltage. A current transformer is connected in series with the power lines to which it is applied so that line current flows in its primary winding. The secondary of the current transformer is connected to current devices such as ammeters, wattmeters, watt-hour meters, power-factor meters, some forms of relays, and the trip coils of some types of circuit breakers.

When no instruments or other devices are connected to the secondary of the current transformer, a short-circuit device or connection is placed across the secondary to prevent the secondary circuit from being opened while the primary is carrying current. If the secondary circuit is opened while the primary winding is carrying current, there will be no secondary ampere turns to balance the primary ampere turns, so the total primary current becomes exciting current and magnetizes the core to a high flux density. This produces a high voltage across both primary and secondary windings and endangers the life of anyone coming in contact with the meters or leads.

SATURABLE-CORE REACTORS

A saturable-core reactor is a magnetic device having a laminated iron core and ac coils similar in construction to a conventional transformer; it is uniquely effective in the control of all types of high-power-factor loads. The coils in a saturable-core reactor are called *gate windings.* In addition, it is designed with an independent winding by which direct current is introduced for control.

When the ac coils of a saturable-core reactor are carrying current to the load, an ac flux (magnetism) saturates the iron core. With only ac coils functioning, the magnetism going into the iron core restricts the ac flow voltage output to the load to about 10 percent of the line supply voltage. Since the iron core is always fully saturated with magnetic flux, the use of direct current from the control winding introduces dc flux, which displaces the ac flux from the iron core. This action reduces the impedance and causes the voltage output to the load to increase. By adjusting the dc flux saturation, the impedance of the ac gate windings may be infinitely varied. This provides a smooth control ranging from approximately 10 to 94 percent of the line voltage at the load, as can be seen in the graph in Fig. 7-6. Since there is practically no power loss in the control of the impedance in a saturable-core reactor, a relatively small amount of direct current can control large amounts of alternating current.

Saturable-core reactors eliminate the need for mechanical and resistance controls and are a very efficient means of proportional power control for resistance heating devices, vacuum furnaces, infrared ovens, process heaters, and other current-limiting applications. In lighting control, especially where wattage per circuit is large, a saturable-core reactor eliminates the loss of dissipated power of a resistance control and provides an infinitely smooth regulated power output to the lighting load.

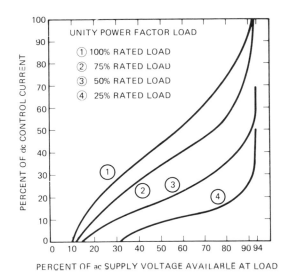

Figure 7-6 By adjusting the dc flux saturation in a saturable-core reactor, the impedance of the ac gate windings may be infinitely varied, as shown in this graph.

Wound-rotor motors can be started smoothly and operate at speeds commensurate with the load when a saturable-core reactor, connected in series with the motor, provides the control. This completely eliminates the maintenance of costly grid resistors and drum controllers.

Although details of design will vary with the manufacturer, usually three basic styles are available: small, medium, and large. The smaller sizes are constructed with two ac coils on a common core, with the dc control winding on the center leg of the core, as shown in Fig. 7-7. Medium-sized reactors normally have two coils, while large-sized reactors normally utilize four coils.

TRANSFORMER GROUNDING

Grounding is necessary to remove static electricity and also as a precautionary measure in case the transformer windings accidentally come in contact with the core or enclosure. All should be grounded and bonded to meet NE Code requirements and also local codes, where applicable.

The tank of every power transformer should be grounded to eliminate the possibility of obtaining static shocks from it or being injured by accidental grounding of the winding to the case. A grounding lug is provided on the base of most transformers for the purpose of grounding the case and fittings.

The NE Code specifically states the requirements of grounding and should be followed in every respect. Furthermore, certain advisory rules recommended by manufacturers provide additional protection beyond that of the

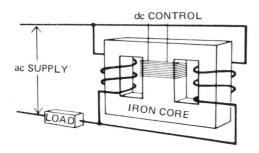

Figure 7-7   The smaller sizes of saturable-core reactors are constructed with two ac coils on a common core with the dc control winding on the center leg of the core.

NE Code. In general, the code requires that separately derived alternating-current systems be grounded as stated in Article 250-26.

For additional information on transformers and their application, the book *Handbook of Power Generation: Transformers and Generators* is recommended. It is authored by John E. Traister and is available from Prentice-Hall, Inc., Englewood Cliffs, New Jersey 07632. This book touches briefly on theory and then progresses immediately into practical applications that can be used for virtually every possible situation.

8

Conductors and Raceways

ELECTROLYTIC COPPER

A variety of materials is used to transmit electrical energy, but copper, due to its excellent cost-to-conductivity ratio, still remains the basic and most ideal conductor. Electrolytic copper, the type used in most electrical conductors, can have three general characteristics:

1. Method of stranding
2. Degree of hardness (temper)
3. Bare, tinned, or coated

Method of Stranding

Stranding refers to the relative flexibility of the conductor and may consist of only one strand or many thousands, depending on the rigidity or flexibility required for a specific need. For example, a small-gauge wire that is to be used in a fixed installation is normally solid (one strand), whereas a wire that will be constantly flexed requires a high degree of flexibility and would contain many strands.

1. Solid is the least flexible form of a conductor and is merely one strand of copper.
2. Stranded refers to more than one strand in a given conductor and may vary from 3 to 19 depending on size.
3. Flexible simply indicates that there are a greater number of strands than are found in normal stranded construction.

Degree of Hardness (Temper)

Temper refers to the relative hardness of the conductor and is noted as soft drawn-annealed (SD), medium hard drawn (MHD), and hard drawn (HD). Again, the specific need of an installation will determine the required temper. Where greater tensile strength is indicated, MHD would be specified over SD, and so on.

Bare, Tinned, or Coated

Untinned copper is plain bare copper that is available in either solid, stranded, or flexible and in the various tempers just described. In this form it is often referred to as *red* copper.

Bare copper is also available with a coating of tin, silver, or nickel to facilitate soldering, to impede corrosion, and to prevent adhesion of the copper conductor to rubber or other types of conductor insulation. The various coatings will also affect the electrical characteristics of copper.

CONDUCTOR SIZE

The American Wire Gauge (AWG) is used in the United States to identify the sizes of wire and cable up to and including No. 4/0 (0000), which is commonly pronounced in the electrical trade as "four-aught" or "four-naught." These numbers run in reverse order as to size; that is, No. 14 AWG is smaller than No. 12 AWG and so on up to size No. 1 AWG. To this size (No. 1 AWG), the larger the gauge number, the smaller the size of the conductor. However, the next larger size after No. 1 AWG is No. 1/0 AWG, then 2/0 AWG, 3/0 AWG, and 4/0 AWG. At this point, the AWG designations end and the larger sizes of conductors are identified by circular mils (CM). From this point, the larger the size of wire, the larger the number of circular mils. For example, 300,000 CM is larger than 250,000 CM. In writing these sizes in circular mils, the "thousand" decimal is replaced by the letter M, and instead of writing, say, 500,000 CM, it is usually written 500 MCM.

ALTERNATE CONDUCTORS

Certain applications in the electrical industry demanded that the characteristics of bare copper be improved on. For example, it was discovered that the maximum operating temperature of copper could be increased by applying a coating of silver or nickel, increasing the temperature rating from 150°C to as high as 260°C. Tensile strength was another problem, and a new process was developed to strengthen the conductor without unduly sacrificing its temperature rating.

There are several processes for combining the conductivity and corrosion resistance of copper with the strength of steel. Copperweld^R is the name of one of these developed by the Copperweld Company. A copper coating is fused to a steel core, and the conductor is available in either 30 or 40 percent conductivity relative to copper. It is widely used in coaxial cables and as a supporting messenger for aerially strung cables. Other higher-strength alloys, such as cadmium, chrome, and zirconium-copper are available, but usually only on special order.

Nickel is a relatively poor conductor of electricity. It resists the flow of current. Normally, electrical resistance is considered a disadvantage. This is not the case where the objective is to generate heat. Electrical resistance creates heat and the operation of any heating element (thermocouples, space heaters, rheostats, etc.) depends on that property of the metal to resist the flow of current. For these requirements, nickel or nickel alloys are the "conductors" of choice.

Another application for nickel is in those areas where copper is inadequate for a high-temperature installation. When the normal operating temperature exceeds 200°C, nickel or a nickel alloy is often used as the conductor.

Nichrome, Monel Metal, constantan, Chromel, and Alumel are all alloys of nickel compounded with other metals, tailored for the specific needs of required resistance or desired high temperature, as follows:

Nichrome	=	nickel-chromium
Monel Metal	=	nickel-copper + (5% of other metals)
Constantan	=	copper-nickel
Chromel	=	nickel-chromium + (5% iron)
Alumel	=	nickel-aluminum

The use of aluminum in small-gauge wires is greatly limited. Its lower cost is an unimportant factor when weighed against its disadvantages:

1. Sixty-one percent of the conductivity of copper
2. Low tensile strength
3. Difficult to solder

Only its light weight can justify its use in limited applications.

Although they are excellent conductors, the high cost of such metals as silver and platinum make their use normally prohibitive. There are applications, however, where their use is justified either in miniaturized circuits or as platings for conductors used in extremely high temperature rated cables.

INSULATIONS

Resistance is that property of a substance that opposes the flow of electricity and is measured in units of *ohms.* The higher the resistance, the more ohms. Resistance is determined by many factors, including the size, length, and type of metal used as the conductor.

Table 8-1 is based on the resistance to electrical flow of a 1-foot length of conductive material with a cross-sectional area of 1 circular mil, expressed in ohms per mil-foot. Expressed another way, the lower the resistance value in the table, the more efficient the conductivity.

TABLE 8-1

Metal	Ohms (per mil ft)
Silver	9.90
Copper	10.37
Gold	14.00
Aluminum	17.00
Tungsten	33.00
Brass	42.00
Cast iron	54.00
Lead	132.00
Constantan	295.00
Mercury	577.00
Nichrome	602.00

The preceding should give us sufficient background for a discussion of insulations, those protective coverings over a conductor that enable both small and large quantities of potentially dangerous electrical current to flow safely and efficiently to where desired.

A typical sample of an electronic cable has four basic components.

1. Conductor
2. Insulation
3. Shield
4. Protective covering or sheath

Insulations for wire and cable fall into two broad categories: (1) electrical and (2) mechanical.

Four major processes are utilized to apply insulation, depending on the characteristics of the insulating material used.

1. Taping
2. Felting
3. Braiding
4. Extruding

A shield is normally a metallic cylinder applied over an insulated conductor or group of conductors.

FUNCTIONS OF SHIELDS

The functions of a shield are many; some are electrical, some mechanical.

1. *Electrical:* To confine the electrical field within the cable or cable system.
2. *Mechanical:* An added bonus of shielding a cable is the additional mechanical protection it affords. This can range from the minimum avoidance of mechanical abuse provided by a foil shield, to the security of a bimetallic corrugated shield used in direct-burial telephone cables.

ENGINEERING CONSIDERATIONS

Many engineering considerations enter into the design of a custom cable. The following information is required to fulfill the electrical requirements of a given specification for custom wire and cable.

1. Current in amperes (milliamperes)
2. Voltage
3. Ohms (dc resistance in megohms/1000 feet)
4. Frequency in hertz (Hz)
5. Capacitance in picofarads (pF)
6. Attenuation in decibels (dB/100 feet)

Capacitance

Capacitance is the *ability of a device to store an electrical charge* or a property of an insulation separating two conductors. This configuration, whether in a cable or an electronic device, is, in fact, a capacitor, and the measure of capacitance is related to the number of electrons that a given voltage can force into storage. This is an important factor in cable selection, and the specifying engineer will usually state the maximum capacitance per foot required in a cable. The unit of capacitance most often used in electronics is the picofarad (pF).

Shielding

Apart from its importance as an electrical barrier, shielding performs a mechanical function as well. From the minimum protection afforded by a foil type shield, to the very effective barrier against water and rodents provided by a corrugated 10-mil copper shield of a direct-burial telephone cable, shielding can figure into the mechanical requirements of a cable.

Jacketing

The choice of jacketing material depends on ambience. Will the cable be installed internally, within a device, overhead outdoors, in conduit or directly buried? Will the environment be extremely hot or extremely cold? Will the cable be subject to a corrosive atmosphere?

Armoring

Armoring is indicated where any wire or cable will be subjected to severe mechanical abuse. The type of armor, however, should be determined by the requirements of the installations. Wire armor is used for vertical drops, interlocked armor is used for defense against extreme mechanical and chemical abuse, and is used for installations requiring maximum moisture and chemical protection.

The selection and application of electrical conductors is based primarily on size of the wires and the type of insulation or covering. A number of tables in Article 310 of the NE Code provide a ready and practical guide to the use of the many types of available conductors in electrical systems.

RACEWAYS

A raceway is any channel used for holding wires, cables, or busbars, which is designed and used solely for this purpose. Types of raceways include rigid metal conduit, intermediate metal conduit (IMC), rigid nonmetallic conduit, flexible metal conduit, liquid-tight flexible metal conduit, electrical metallic tubing (EMT), underfloor raceways, cellular metal floor raceways, cellular concrete floor raceways, surface metal raceways, wireways, and auxiliary gutters. Raceways are constructed of either metal or insulating material.

Raceways provide mechanical protection for the conductors that run in them and also prevent accidental damage to insulation and the conducting metal. They also protect conductors from the harmful chemical attack of corrosive atmospheres and prevent fire hazards to life and property by confining arcs and flame due to faults in the wiring system.

One of the most important functions of metal raceways is to provide a path for the flow of fault current to ground, thereby preventing voltage buildups on conductor and equipment enclosures. This feature, of course, helps to minimize shock hazards to personnel. To maintain this feature, it is extremely important that all raceway systems be securely bonded together into a continuous conductive path and properly connected to a grounding electrode such as a water pipe or a ground rod.

The NE Code provides rules on the maximum number of conductors permitted in raceways. In conduits, for either new work or rewiring of existing raceways, the maximum fill must not exceed 40 percent of the conduit cross-sectional area. In all such cases, fill is based on using the actual cross-sectional areas of the particular types of conductors used. Other derating rules specified by the NE Code may be found in Article 310. For example, if more than three conductors are used in a single conduit, a reduction in carrying capacity is required. Ambient temperature is another consideration that may call for derating of wires below the values given in NE Code, Tables 310-6 through 310-19.

9

Branch Circuits and Feeders

The conductors that extend from the panelboard to the various outlets are called branch circuits and are defined by the NE Code as "that point of a wiring system extending beyond the final overcurrent device protecting the circuit . . . ."

In general, the size of the branch-circuit conductors varies depending on the current requirements of the electrical equipment connected to the outlet. Most, however, will consist of either No. 14, 12, 10, or 8 AWG. Conductors larger than No. 8 AWG are usually considered to be *feeders* rather than branch circuits.

A simple branch circuit requires two wires or conductors to provide a continuous path for the flow of electric current. The usual branch circuit for receptacles and lighting operates at 120 volts, although many lighting circuits in commercial occupancies operate at 277 volts.

Fractional-horsepower motors and small electric heaters usually operate at 120 volts also and are connected to a simple 120-volt branch circuit either by means of a receptacle, a junction box, or a direct connection. Larger electric motors, air-conditioning, duct and unit heaters, and other large current-consuming equipment operate on a two- or three-wire circuit at 240 or 480 volts.

The NE Code specifically states that the total load on a branch circuit, other than motor loads, must not exceed 80 percent of the circuit rating when the load will constitute a continuous load, that is, a load that is in operation for 3 hours or longer.

To size the total load permitted on a circuit of, say, 20 amperes consisting of No. 12 AWG copper wire and fused at 20 amperes that will be in operation for 3 hours or longer, multiply the amperes (20) by 0.80 to obtain 16 amperes. This figure multiplied by the voltage (120) gives a maximum of 1920 watts that may be connected to the circuit. However, on new wiring systems it is best in most cases to limit the load on 20-ampere circuits to a maximum of 1200 to 1600 watts. This permits some loads to be added in the future, and the practice also keeps the temperature of the conductors lower for better performance and less voltage drop.

SIZING CONDUCTORS

In all electrical systems, the conductors should be sized so that the voltage drop never exceeds 3 percent for power, heating, and lighting loads or combinations of these. Furthermore, the maximum total voltage drop for conductors for feeders and branch circuits combined should never exceed 5 percent.

The voltage drop in any two-wire, single-phase circuit consisting of mostly resistance-type loads, with the inductance negligible, may be found by the following equation:

$$VD = \frac{2K \times L \times I}{CM}$$

where

VD = drop in circuit voltage
L = length of conductor
I = current in the circuit
CM = area of conductor in circular mils
K = resistivity of conductor metal, that is, 11 for copper and 18 for aluminum

With this equation, the voltage drop in a circuit consisting of No. 10 AWG copper wire, 50 feet in length, and carrying a load of 20 amperes would be

$$VD = \frac{2(11) \times 50 \times 20}{10,380 \text{ (area in CM of No. 10 wire)}}$$

which equals 2.12 volts. Divided by 120, this is .01 or 1 percent, well within the limits of the allowed voltage drop.

It is also quite common to deal with the problem of sizing conductors for special electronic and electrical equipment on which conventional wiring sizing equations and calculations cannot be used. For example, x-ray and computer equipment are highly sensitive to voltage variations, and the circuits must be sized to obtain the very minimum of voltage drop.

To illustrate, assume that a solid-state computer is located 460 feet from the main distribution panelboard and it is necessary to keep the voltage drop less than the 3 percent allowable. The service consists of a three-phase, four-wire, 120/208-volt circuit, and the allowable voltage drop is 2 percent. Copper wire will be utilized.

To find the wire size that will carry the load with less than 2 percent voltage drop, use the equation

$$\text{Circular mils} = \frac{\text{length} \times \text{amperes} \times 2K}{\text{volts lost}}$$

In this equation the number of feet must be measured or scaled one way, not both sides of the circuit; volts lost should be taken as the drop allowed in volts, not the percentage. Circular mils show the size of wire in AWG to use, while $K = 11$ for copper.

There are only two unknowns in this situation: the circular mils and the volts lost. To solve for the voltage drop or volts lost, multiply 208 (the circuit voltage between phases) by 0.02 (the percentage of voltage drop permitted for the particular piece of equipment). The answer will be 4.16 volts. The nameplate on the piece of equipment gives a full-load ampere rating of 87 amperes at 208 volts. Substituting all known values in the equation, we have

$$\text{Circular mils} = \frac{460 \times 87 \times 22}{4.16} = 211{,}644 \text{ CM}$$

Referring to a wire size table, 250,000 CM (250 MCM) is the closest wire size normally available and will therefore be the size to use for the circuit feeding the piece of equipment.

Most electricians and designers never use conductors smaller than No. 12 AWG on branch circuits for power and lighting to help cut down on the voltage drop, which wastes energy, although No. 14 AWG is permitted for use on branch circuits.

10

Wire Connections and Splices

Anyone who is involved in the installation of electrical systems should have a good knowledge of wire connectors and splicing, as it is necessary to make numerous electrical joints during the course of any electrical installation.

Splices and connections that are properly made will often last as long as the insulation on the wire itself, while poorly made connections will always be a source of trouble; that is, the joints will overheat under load and cause a higher resistance in the circuit than should be.

The basic requirements for a good electrical connection include the following:

1. It should be mechanically and electrically secure.
2. It should be insulated as well as or better than the existing insulation on the conductors.
3. These characteristics should last as long as the conductor is in service.

There are many different types of electrical joints for different purposes, and the selection of the proper type for a given application will depend to a great extent on how and where the splice or connection is used.

Currently, electrical joints are normally made with solderless pressure connectors to save time, but electrical workers should also have a knowledge of the traditional splices.

STRIPPING AND CLEANING CONDUCTORS

Before any connection or splice can be made, the ends of the conductors must be properly stripped and cleaned. Stripping is the removal of insulation from the conductors at the end of the wire or at the location of the splice. Experienced electricians can strip wire very well with a pocket knife or a pair of side-cutting pliers (for the smaller sizes), but there are many handy tools on the market that will facilitate this operation. The use of such tools will also help to prevent cuts and nicks in the wire, which reduce the conductor area as well as weaken the conductor.

Poorly stripped wire can result in nicks, scrapes, or burnishes. Any of these can lead to a stress concentration at the damaged cross section. Heat, rapid temperature change, mechanical vibration, and oscillatory motion can aggravate the damage, causing improper signals in the circuitry or even total failure.

Lost strands are a problem in splices or crimp-type terminals, while exposed strands might be a safety hazard.

Slight burnishes on conductors, as long as they had no sharp edges, were acceptable at one time. Now, however, reliability experts feel that under certain conditions removing as little as 40 microinches of conductor plating from some wires can cause failure.

Faulty stripping can pierce, scuff, or split the insulation. This can cause changes in dielectric strength and lower the wire's resistance to moisture and abrasion. Insulation particles often get trapped in solder and crimp joints. These form the basis for a defective termination. A variety of factors determine just how precisely a wire can be stripped: wire size, insulation concentricity, adherence, and others.

It is a common mistake to believe that a certain gauge of stranded conductor has the same diameter as a solid conductor. This is a very important consideration in selecting proper blades or strippers. Figure 10-1 shows the nominal sizes referenced for the different gauges.

The type of stripping method chosen for the job is very important. Specifications and standards, both commercial and military, vary considerably on these points.

WIRE CONNECTIONS UNDER 600 VOLTS

Wire connections are used to connect a wire to an electrical device such as a receptacle, wall switch, or pump-control switch.

Electricians will normally encounter wiring devices with screw terminals more often than any other type, and the simple *eye* connection is the one to use for such terminals. To make the eye in the wire, first strip and clean approximately 1 to 1-1/2 inches of insulation from the end of the wire. With

AWG	Dia. of Solid Wire (inches)	Dia. of Stranded Wire (inches)	Dia. of Solid Wire (mm)	Dia. of Stranded Wire (mm)
8	0.128	0.145 to 0.149	3.512	3.65 to 4.29
10	0.102	0.116 to 0.119	2.590	2.59 to 3.05
12	0.081	0.091 to 0.093	2.057	2.13 to 2.14
14	0.064	0.072 to 0.074	1.625	1.67 to 1.90
16	0.051	0.058 to 0.060	1.295	1.35 to 1.52
18	0.040	0.047 to 0.049	1.016	1.09 to 1.29
20	0.032	0.038 to 0.040	0.813	0.965 to 1.05
22	0.025	0.029 to 0.030	0.635	0.736 to 0.838
24	0.020	0.024 to 0.025	0.508	0.590 to 0.670
26	0.016	0.019 to 0.020	0.406	0.482 to 0.540
28	0.013	0.016 to 0.017	0.330	0.360 to 0.432
30	0.010	0.012 to 0.013	0.254	0.305 to 0.357

Figure 10-1   Normal wire sizes for different gauges.

a pair of long-nose pliers, make a slight bend in the wire near the insulation and at an angle of approximately 45°. Continue by bending the wire (above the first bend) in the opposite direction and at different points to form a circle in the wire. The eye may then be placed under the screw head so that the end follows in the direction in which the screw will be tightened; this will cause the eye to close tightly around the screw threads. If the eye is reversed, it will open and be loose around the threads.

Other types of wire connectors are shown in Fig. 10-2. These connectors are available in various sizes to accommodate wire from No. 22 AWG through 250 MCM. They can be installed with crimping tools having a single indenter or double indenter. Wire range identification is normally stamped on the tongue of each terminal.

Terminators are also available in larger sizes to accommodate wires from No. 8 AWG through 1000 MCM. Two types, one-hole lugs and two-way connectors, are shown in Fig. 10-3.

CONNECTING ALUMINUM CONDUCTORS

Aluminum has certain properties that are different from copper that must be understood if reliable connections are to be made. These properties are: cold flow, coefficient of thermal expansion, susceptibility to galvanic corrosion, and the formation of oxide film on the metal's surface.

Because of thermal expansion and cold flow of aluminum, standard copper connectors as found on the market today cannot be safely used on aluminum wire. Most manufacturers design their aluminum connectors with greater contact areas to counteract this property of aluminum. Tongues and barrels of all aluminum connectors are larger or deeper than comparable copper connectors.

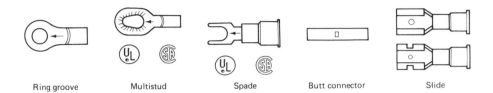

Ring groove   Multistud   Spade   Butt connector   Slide

Copper Mechanical Lugs and Connectors

· U.L. listed
· C.S.A. certified
· Heavy-duty seamless construction

One barrel, offset tongue,
one hole, Type CB
No. 14 AWG through 1000 MCM

One barrel, offset tongue,
two hole, Type CO
No. 14 AWG through 1000 MCM

One barrel, fixed tongue,
one hole, Type CX
No. 14 AWG through 500 MCM

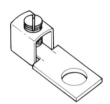

One barrel, straight tongue,
one hole, Type CS
No. 14 AWG through 1000 MCM

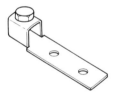

One barrel, straight tongue,
two hole, Type CD
No. 14 AWG through 1000 MCM

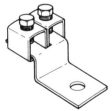

Two barrels, offset tongue,
one hole, Type DC
No. 6 AWG through 500 MCM

Split bolt connector, Type SBC
(2) No. 14 AWG through (2) 1000 MCM
run and tap combinations

Figure 10-2   Type of wire connectors in current use.

Figure 10-3
One hole and two-way
connectors.

The electrolytic action between aluminum and copper can be controlled by plating the aluminum with a neutral metal (usually tin). The plating prevents electrolysis from taking place and the joint remains tight. As an additional precaution, a joint sealing compound should be used. Connectors should be tin plated and prefilled with an oxide abrading sealing compound.

The insulating aluminum oxide film must be removed or penetrated before a reliable aluminum joint can be made. Aluminum connectors are designed to bite through this film as they are applied to conductors. It is further recommended that the conductor be wire brushed and preferably coated with a joint compound to guarantee a reliable joint.

Connectors marked with just the wire size should only be used with copper conductors. Connectors marked with Al and the wire size should only be used with aluminum wire. Connectors marked with Al-Cu and the wire size may be safely used with either copper or aluminum.

TAPING ELECTRICAL WIRE JOINTS

All wire joints not protected by some other means should be taped carefully to provide the same quality of insulation over the splice as over the rest of the wires. Spliced joints in rubber-insulated wires should be covered with rubber and friction tape, while joints in thermoplastic-insulated wires should be covered with pressure-sensitive thermoplastic-adhesive tape such as Scotch No. 33 for indoor use or Scotch No. 88 for outside use.

For covering rubber-insulated wires, apply the rubber tape to the splice first to provide air- and moisture-tight insulation. The amount applied should be equal to the insulation that was removed. Then wrap the friction tape over the rubber tape to provide mechanical protection. For joints in thermoplastic insulated wires, just the one tape is enough, that is, Scotch No. 33 or No. 88.

To start the taping of a splice or joint, start the end of the tape at one end of the splice (see Fig. 10-4), slightly overlapping the insulation on the wires. Stretch it slightly while winding it on in a spiral. When the joint is completely covered with layers equal to the original insulation, press or pinch the end of the tape down tightly onto the last turn to make it stick. A properly wrapped electrical joint is shown in Fig. 10-5.

Other splicing and taping techniques are shown in Fig. 10-6. Figure 10-7 shows the method of insulating a split-bolt connector with filler tape.

Figure 10-4   Start the end of the tape at one end of the splice.

Figure 10-5   Properly wrapped electrical joint.

1. Cut a piece of Slipknot Filler Tape and place over each side of the connector.

2. Wrap both pieces around the connector, using moderate finger pressure to shape the filler tape.

3. Wrap the covered connector with Slipknot Plastic Tape.

Figure 10-6   Method of insulating a split bolt connector.

HIGH-VOLTAGE SPLICES AND CONNECTIONS

High-voltage splicing is an art all to itself and only those specially trained should attempt such work. However, a brief explanation of the technique is in order. These techniques are generalized for illustrative purposes. For specific splices, the reader may contact the Tape Division of the Plymouth Rubber Company, Canton, Massachusetts 02021 or the cable manufacturer for application procedures.

For a typical straight splice for a single (unshielded) 600- to 5000-volt conductor, the jacket and insulation should be removed one half of the connector length plus 1/4 inch. Apply a protective wrap of friction tape to the jacket ends (Fig. 10-8) to protect the jacket and insulation while cleaning the conductor and securing the connector.

After the preceding operations are performed, remove the friction tape and also remove additional jacket and insulation in an amount equal to 25 times the thickness of the overall insulation, extending in either direction

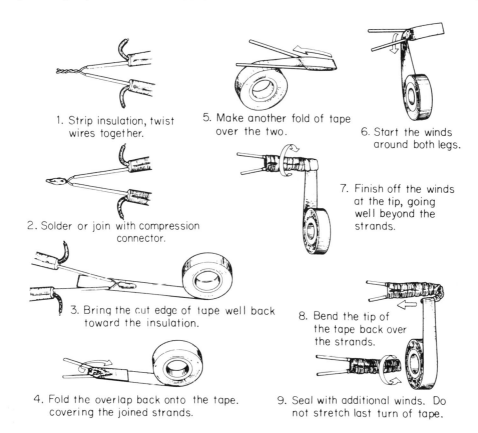

1. Strip insulation, twist wires together.

2. Solder or join with compression connector.

3. Bring the cut edge of tape well back toward the insulation.

4. Fold the overlap back onto the tape, covering the joined strands.

5. Make another fold of tape over the two.

6. Start the winds around both legs.

7. Finish off the winds at the tip, going well beyond the strands.

8. Bend the tip of the tape back over the strands.

9. Seal with additional winds. Do not stretch last turn of tape.

Figure 10-7   Typical low-voltage splicing methods.

from the first insulation cut. For example, if the thickness of the insulation is 1/4 inch, the outside edge of the splice will be 6-1/4 inches beyond the insulation cut. Pencil insulation with a sharp knife, taking care not to nick the conductor.

Refer to Fig. 10-9 as you read the following instructions. Starting from A to B, wrap the splice area with Plymozone, the ozone-resistant splicing compound, stretching the rubber tape about 25 percent and unwinding the Holland cloth interliner as you wrap. The finished splice should be built up one and one-half times the factory insulation if rubber (two times if thermoplastic) measured from the top of the connector. If the factory insulation is 1/4 inch, the finished splice would be 3/8-inch thick when measured from the top of the connector and should extend about 1-1/2 inches beyond the jacket edge. To finish off the splice and to ensure complete protection, cover the entire splice with four layers of Slipknot Friction or 6 layers of plastic tape, extending about 1-1/2 inches beyond the edge of the splicing compound.

For shielded cable, follow the cable manufacturer's recommendations for securing the shielding system. Then top off with plastic tape.

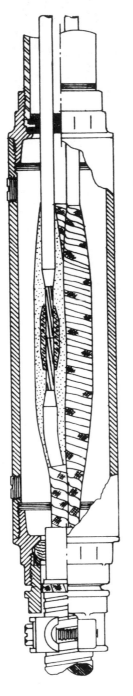

Figure 10-8 Apply a protective wrap of friction tape to the jacket ends to protect the jacket and insulation while cleaning the conductor and securing the connector.

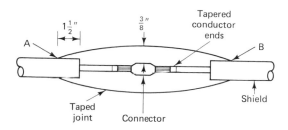

Figure 10-9   Method of insulating and taping the splice.

HEAT-SHRINK INSULATORS

Heat-shrinkable insulators for small connectors provide skintight insulation protection and are fast and easy to use. They are designed to slip over wires, taper pins, connectors, terminals, and splices. When heat is applied, the insulation becomes semirigid and will provide positive strain relief at the flex point of the conductor. A vaporproof band will seal and protect the conductor from abrasion, chemicals, dust, gasoline, oil, and moisture. Extreme temperatures, both hot and cold, will not affect the performance of these insulators. The source of heat can be any numbers of types, but most manufacturers of these insulators also produce a heat gun especially designed for use on heat-shrink insulators. It closely resembles and operates the same as a conventional hair dryer with fan blower (see Figs. 10-10 and 10-11).

In general, a heat-shrink insulator may be thought of as tubing with a memory. When initially manufactured, it is heated and expanded to a predetermined diameter and then cooled. Upon application of heat, through various methods, the tubing compound "remembers" its original size and shrinks to that smaller diameter. It is available in a range of sizes and is designed to shrink easily over any wire or device when heat is applied. This property enables it to conform to the contours of any object. The following describe some of the types currently available.

PVC: This type is a general-purpose, economical tubing that is widely used in the electronic industry. The PVC compound is irradiated by being bombarded with high-velocity electrons. This results in a denser, cross-linked material with superior electrical and mechanical properties. It also ensures that the tubing will resist cracking and splitting.

Slip insulator over object to be insulated. Apply heat for a few seconds.

Finished... Permanent Insulation Protection.

Figure 10-10   Method of installing heat-shrink insulators. (Courtesy Thomas & Betts.)

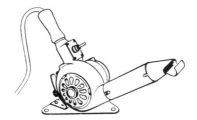

Figure 10-11   Typical heat gun
for use with heat shrink insulators.
(Courtesy Thomas & Betts.)

Polyolefin: Polyolefin tubing has a wide range of uses for wire bundling, harnessing, strain relief, and other applications where cables and components require additional insulation. It is irradiated, flame retarding, flexible, and comes in a wide variety of colors.

Double wall: This type is available and designed for outstanding protective characteristics. It is semirigid tubing that has an inner wall that actually melts and an outer wall that shrinks to conform to the melted area.

Teflon^R :   This type is considered by many users to be the best overall heat-shrinkable tubing—physically, electrically, and chemically. Its high temperature rating of 250°C resists brittleness or loss of translucency from extended exposure to high heat and will not support combustion.

Neoprene: Components that warrant extra protection from abrasion require a highly durable, yet flexible, tubing. The irradiated neoprene tubing offers this optimal coverage.

Kynar^R: Irradiated Kynar is a thin-wall, semirigid tubing with outstanding resistance to abrasion. This transparent tubing enables easy inspection of components that are covered and retains its properties at its rated temperature.

Most tubing is available in a wide variety of colors and put-ups. The Tubing Selector Guide shown in Fig. 10-12 was compiled by the Manhattan Electric Cable Corporation and can help in the selection of the best tubing for any given application.

Manhattan Number	Type	Material	Temperature Range, °C	Shrink Ratio	Max. Long. Shrinkage, %	Tensile Strength, psi	Colors	Dielectric Strength, V/mil
MT-105	Nonshrinkable	PVC	+105	–	–	2700	White, red, clear, black	800
MT-150	Shrinkable	PVC	−35 to +105	2:1	10	2700	Clear, black	750
MT-200	Nonshrinkable	Teflon[R]	−65 to +260	–	–	2700	Clear	1400
MT-221	Shrinkable	Flexible polyolefin	−55 to +135	2:1	5	2500	Black, white, red, yellow, blue, clear	1300
MT-250	Nonshrinkable	Teflon[R]	−65 to +260	–	–	7500	Clear	1400
MT-300	Shrinkable	Polyolefin double wall	−55 to +110	6:1	5	2500	Black	1100
MT-350	Shrinkable	Kynar[R]	−55 to +175	2:1	10	8000	Clear	1500
MT-400	Shrinkable	Teflon[R]	+250	1.2:2	10	6000	Clear	1500
MT-500	Shrinkable	Teflon[R]	+250	11/2:1	10	6000	Clear	1500
MT-600	Shrinkable	Neoprene	+120	2:1	10	1500	Black	300

Figure 10-12 Tubing selector guide. (Courtesy Manhattan Electric Cable Corp.)

11

Grounding

All building electrical systems must be grounded in a manner prescribed by the NE Code to protect personnel and valuable equipment. To be totally effective, a grounding system must accomplish the following:

1. Provide a low-impedance path to ground for personnel and equipment protection and effective circuit relaying.
2. Withstand and dissipate repeated fault and surge circuits.
3. Provide corrosion resistance to various soil chemistries to ensure continuous performance for the life of the equipment being protected.
4. Provide rugged mechanical properties for easy driving with minimum effort and rod damage.

OSHA AND NE CODE REQUIREMENTS

The grounding equipment requirements established by Underwriters' Laboratories, Inc., has served as the basis for approval for grounding of the NE Code. The NE Code, in turn, provides the grounding premises of the Occupational Safety and Health Act (OSHA, 1971).

In general, the electrical system is grounded at the service-entrance location. This is accomplished by using a grounding conductor connected to a driven ground rod and a cold-water pipe (see Figs. 11-1 and 11-2). The opposite end of the grounding conductor is connected to the service neutral conductor usually by means of a solid neutral block, with a jumper conductor connected between the neutral block and the service grounding conductor.

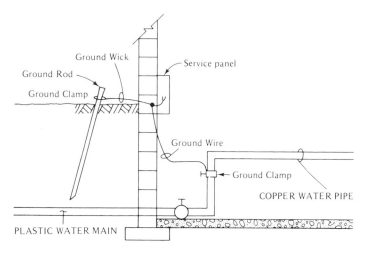

Figure 11-1   One method of grounding a service entrance.
(Courtesy Sorgel Transformers Square D Co.)

The current edition of the NE Code states that a metal underground water pipe, previously preferred as a system ground, shall be supplemented by an additional electrode. The type of grounding electrode generally considered most convenient is the copper-bonded ground rod.

The NE Code and OSHA require that a copper-bonded ground rod shall have a minimum diameter of 1/2 inch and shall also be at least 8 feet in length. It further stipulates that the ground rod resistance should not exceed 25 ohms.

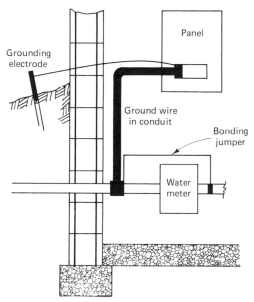

Figure 11-2   Another method of grounding a service entrance. (Courtesy Sorgel Transformers Square D Co.)

While the NE Code considers the metallic enclosures of the service-entrance equipment to be adequately grounded when the service-entrance raceway is mechanically connected to it, some area local ordinances require that a grounding-conductor jumper also be installed from the service raceway-grounding clamp to the service-entrance-equipment metal enclosure, as shown in Fig. 11-3.

Branch circuits and the metal enclosures of panelboards are considered to be adequately grounded when they are mechanically connected (bonded)

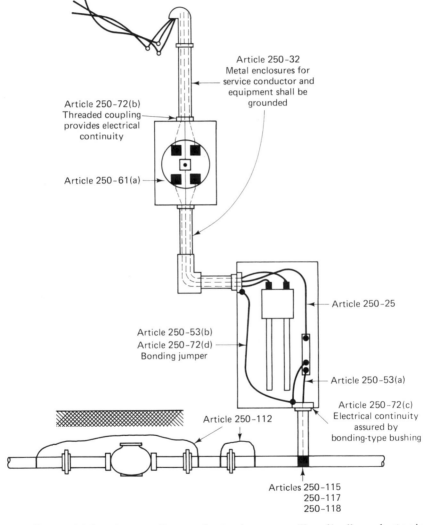

Figure 11-3 A grounding-conductor jumper or "bonding" conductor is required to ensure continuity on some electrical systems in certain areas. (Courtesy Sorgel Transformers Square D Co.)

to each other and to the service-entrance-equipment metal enclosure. This is accomplished either by metallic raceways (conduit) or by a ground wire enclosed within nonmetallic cable.

Circuits are grounded to limit the voltage on the circuit and protect it from the following:

1. Exposure to lightning.
2. Voltage surges higher than that for which the circuit is designed.
3. An increase in the maximum potential to ground due to normal voltage.

On a 120/240-volt, single-phase power system supply, such as that used to service most residences, the neutral should be grounded at the transformer ahead of the main-service disconnecting means (see Fig. 11-4). Another ground is normally required at the electric meter or at the main panelboard. This is accomplished by connecting a grounding conductor to the neutral bus of the panel or meter base and by connecting the other end to a cold-water pipe and also a suitable grounding electrode. No further connection should be made to a grounding electrode on the load side of the main-service switch.

The conductor that is grounded (neutral) should be white in all cases. If white insulation is not available, such as on the larger-sized conductors, the conductor must be identified at each connection, junction box, or pull

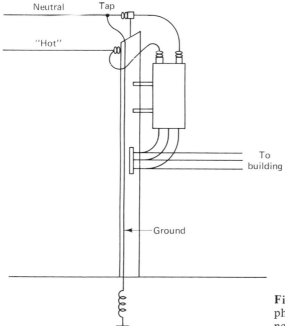

Figure 11-4 A 120/240-volt, single-phase transformer grounded at the neutral ahead of the main-service disconnecting means.

box by either white paint or an approved white tape secured around the neutral conductor.

Do not confuse the grounded conductor with the grounding conductor. As the names imply, the neutral is a conductor that is grounded and should be white in all cases. On the other hand, the grounding conductor is a conductor used for grounding; this conductor may be bare copper or some other noncorrosive conducting material such as aluminum. If it is insulated, the color must be green or green with a yellow tracer.

A good grounding system must meet certain criteria:

1. Provide a low resistance path to ground.
2. Withstand high impulse and fault currents.
3. Resist corrosion while in contact with all types of soil.
4. Withstand mechanical impact and abrasion by foreign objects while being installed.

The other half of the ground connecting system is the earth itself. First, the area of contact between the earth and ground rod must be sufficient so that the resistance of the current path into and through the earth will be within the allowable limits of the particular application. The resistance of this earth path must be relatively low and must remain reasonably constant throughout the seasons of the year.

To understand why earth resistance must be low, see Fig. 11-5 and then apply Ohm's law, $E = I \times R$. (E is volts, I is the current in amperes, and R is the resistance in ohms.) For example, assume a 4000-volt supply (2300 volts to "ground"), with a resistance of 13 ohms. Now assume an exposed wire in

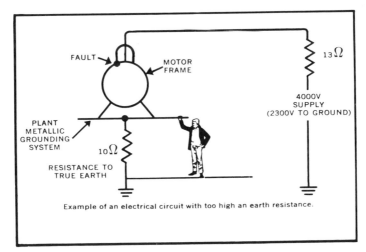

Example of an electrical circuit with too high an earth resistance.

Figure 11-5   An electric circuit with too high an earth resistance. (Courtesy ITT Weaver.)

this system touches a motor frame that is connected to a grounding system that has a 10-ohm resistance to earth.

According to Ohm's law, there will be a current of 100 amperes through the fault, from the motor frame to the earth. If a person touches the motor frame and is solidly grounded to earth, he could be subjected to 1000 volts (10 ohms times 100 amperes). This is more than enough for a fatality.

NATURE OF AN EARTH ELECTRODE

Resistance to current through an earth electrode system has three components:

1. Resistance of the ground rod itself and connections to it.
2. Contact resistance between the ground rod and the earth adjacent to it.
3. Resistance of the surrounding earth.

Ground rods, masses of metal, structures, and other devices are commonly used for earth electrodes. These electrodes are usually adequate in size or cross section so that their resistance is a negligible part of the total resistance.

The resistance between the electrode and the earth is much less than normally suspected. If the electrode is free of paint or grease, and the earth is firmly packed, the U.S. National Bureau of Standards No. 108 has shown that contact resistance is negligible.

A ground rod driven into earth of uniform resistivity conducts current in all directions. Think of the electrode as being surrounded by shells of earth, all of equal thickness (see Fig. 11-6). The earth shell closest to the ground rod has the smallest surface area and consequently offers the greatest resistance. The next earth shell is somewhat larger in area and offers less resistance. Finally, a distance from the ground rod will be reached where

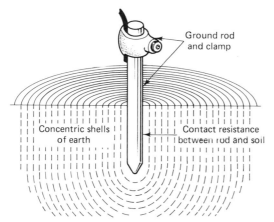

Figure 11-6   Resistive components of earth electrode. (Courtesy ITT Weaver.)

inclusion of additional earth shells does not add significantly to the "earth resistance" surrounding the ground rod. This distance is approximately 8 to 10 feet, and is known as the *effective resistance area;* it is mainly dependent on the depth of the ground rod.

Of the three components involved in resistance, the resistivity of the earth is most critical and most difficult to calculate and overcome.

CALCULATING EARTH RESISTANCE

The equations for earth resistance from various systems of electrodes are quite complicated and in some cases may be expressed only as approximations. All such formulas are derived from the general formula $R = pL/A$ and are based on the assumption of uniform earth resistivity throughout the entire soil volume being considered, although this is seldom the case.

The following is a commonly used resistance-to-earth formula for single ground rods developed by H. R. Dwight of the Massachusetts Institute of Technology:

$$R = \frac{p}{2\mu L} \ln \frac{4L}{a} - 1$$

where

       p = average soil resistivity, ohms-centimeters
       L = ground rod length, centimeters
       a = ground rod radius, centimeters
       R = resistance of rod to earth, ohms

EFFECT OF ROD SIZE ON RESISTANCE

Whenever a ground rod is driven deeper into the earth, its resistance is substantially reduced. Generally, doubling the rod length reduces resistance by an additional 40 percent. Increasing the diameter of the rod, however, does not materially reduce its resistance. Doubling the diameter, for instance, reduces resistance by less than 10 percent.

EFFECTS OF SOIL RESISTIVITY ON EARTH ELECTRODE RESISTANCE

Dwight's formula shows that the resistance to earth of grounding electrodes depends not only on the depth and surface area of grounding electrodes but on soil resistivity as well.

Soil resistivity is the key factor that determines what the resistance of a grounding electrode will be and to what depth it must be driven to obtain low ground resistance. The resistivity of the soil varies widely throughout the world and changes seasonally. Soil resistivity is determined largely by its electrolyte content, consisting of moisture, minerals, and dissolved salts. A dry soil has high resistivity, but a wet soil may also have a high resistivity if it contains no soluble salts. Some values found for earth resistivity are given in Fig. 11-7.

Soil	Resistivity, Ohm-cm (Range)		
Surface soils, loam, etc...............	100	—	5,000
Clay...................................	200	—	10,000
Sand and gravel......................	5,000	—	100,000
Surface limestone....................	10,000	—	1,000,000
Limestones...........................	500	—	400,000
Shales................................	500	—	10,000
Sandstone............................	2,000	—	200,000
Granites, basalts, etc................	100,000		
Decomposed gneisses................	5,000	—	50,000
Slates, etc...........................	1,000	—	10,000

Figure 11-7   Resistivities of different soils. (Courtesy ITT Weaver.)

FACTORS AFFECTING SOIL RESISTIVITY

Figure 11-8 shows, for two samples of soil, the variation of resistivity with moisture content. These two samples of soil, when thoroughly dried, become in fact very good insulators, having a resistivity in excess of 10^9 ohm-centimeters. The resistivity of the soil sample is seen to change quite rapidly until approximately 20 percent or greater moisture content is reached.

The resistivity of the soil is also influenced by temperature. Figure 11-9 shows the variation of the resistivity of sandy loam containing 15.2 percent moisture, with temperature changes from $20°$ to $-15°$C. In this temperature range the resistivity is seen to vary from 7200 to 330,000 ohm-centimeters.

Moisture Content, % By Weight	Resistivity, Ohm-cm	
	Top Soil	Sandy Loam
0	$1,000 \times 10^6$	$1,000 \times 10^6$
2.5	250,000	150,000
5	165,000	43,000
10	53,000	18,500
15	19,000	10,500
20	12,000	6,300
30	6,400	4,200

Figure 11-8   Effect of moisture content on earth resistivity. (Courtesy ITT Weaver.)

Because soil resistivity directly relates to moisture content and temperature, it is reasonable to assume that the resistance of any grounding system
will vary throughout the different seasons of the year. Such variations are
shown in Fig. 11-10. Since both temperature and moisture content become
more stable at greater distances below the surface of the earth, it follows
that a grounding system, to be most effective at all times, should be constructed with the ground rod driven down a considerable distance below the
surface of the earth. Best results are obtained if the ground rod reaches the
permanent moisture level.

Temperature		Resistivity, Ohm-cm
C	F	
20	68	7,200
10	50	9,900
0	32 (water)	13,800
0	32 (ice)	30,000
−5	23	79,000
−15	14	330,000

Figure 11-9   Effect of temperature on earth resistivity.
(Courtesy ITT Weaver.)

In some locations, the resistivity of the earth is so high that low-
resistance grounding can be obtained only at considerable expense and with
an elaborate grounding system. In such situations, it may be economical to
use a ground rod system of limited size and to reduce the ground resistivity by
periodically increasing the soluble chemical content of the soil. Figure 11-11
shows the substantial reduction in resistivity of sandy loam brought about
by an increase in chemical salt content.

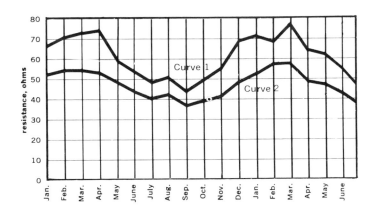

Figure 11-10   Seasonal variation of earth resistance with an electrode
of 3/4-inch pipe in rather stony clay soil. Depth of electrode in earth is
3 feet for curve 1 and 10 feet for curve 2. (Courtesy ITT Weaver.)

Chemically treated soil is also subject to considerable variation of resistivity with changes in temperature, as shown in Fig. 11-12. If salt treatment is employed, it is necessary to use ground rods that will resist chemical corrosion.

To assist the engineer in determining the approximate ground rod depth required to obtain a desired resistance, a device called the grounding nomograph can be used. The nomograph (Fig. 11-13) indicates that to obtain a grounding resistance of 15 ohms in a soil with a resistivity of 15,000 ohm-centimeters, a 5/8-inch rod must be driven 45 feet. It should be noted that the values indicated on the nomograph are based on the assumption that the soil is homogeneous and, therefore, has uniform resistivity.

Under normal conditions, the table in Fig. 11-14 can be used for sizing grounding conductors for residential and small commercial applications. For example, assume that a residence requires a 200-ampere service entrance, and 4/0 aluminum conductors are used for the service-entrance conductors. The table in Fig. 11-14 gives two types of wire to choose from, copper and aluminum. With 4/0 aluminum conductors used for the main service conductors, the table gives size 4 AWG or size 2 AWG for copper and aluminum, respectively.

If the resistance of the ground rod is doubtful, a megohm ground tester can be used to check the resistance. If a metal main-water line is used for the ground source, however, a check is usually not necessary, since the resistance of such sources is nearly always below 3 ohms, well within the NE Code requirements.

**THE EFFECT OF SALT * CONTENT ON THE
RESISTIVITY OF SOIL**
(Sandy loam. Moisture content, 15 per cent by weight.
Temperature, 17°C

Added Salt (Per cent by weight of moisture)	Resistivity (Ohm-centimeters)
0	10,700
0.1	1,800
1.0	460
5	190
10	130
20	100

Figure 11-11   The effect of salt content on the resistivity of soil. (Courtesy ITT Weaver.)

**THE EFFECT OF TEMPERATURE ON THE
RESISTIVITY OF SOIL CONTAINING SALT ***

(Sandy loam, 20 per cent moisture. Salt 5 per cent of weight of moisture)

Temperature (Degrees C)	Resistivity (Ohm-centimeters)
20	110
10	142
0	190
− 5	312
−13	1440

Figure 11-12   The effect of temperature on the resistivity of soil containing salt. (Courtesy ITT Weaver.)

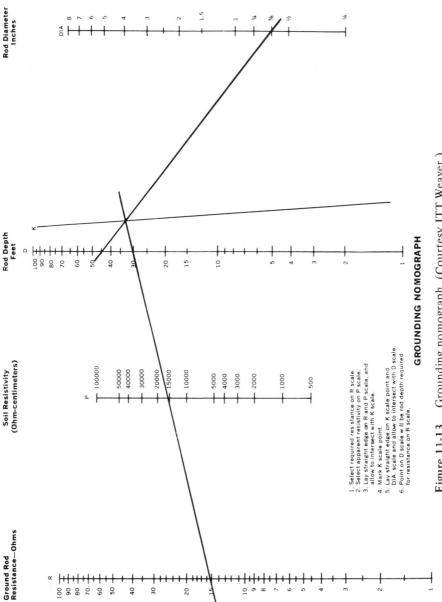

Figure 11-13 Grounding nomograph. (Courtesy ITT Weaver.)

GROUNDING NOMOGRAPH

Rod Diameter Inches

DIA

Rod Depth Feet

D

Soil Resistivity (Ohm–centimeters)

P

Ground Rod Resistance—Ohms

R

K

1. Select required resistance on R scale.
2. Select apparent resistivity on P scale.
3. Lay straight edge on R and P scale, and allow to intersect with K scale.
4. Mark K scale point.
5. Lay straight edge on K scale point and DIA. scale and allow to intersect with D scale.
6. Point on D scale will be rod depth required for resistance on R scale.

Table 250-94   Grounding Electrode Conductor for AC Systems

Size of Largest Service-Entrance Conductor or Equivalent for Parallel Conductors		Size of Grounding Electrode Conductor	
Copper	Aluminum or Copper-Clad Aluminum	Copper	*Aluminum or Copper-Clad Aluminum
2 or smaller	0 or smaller	8	6
1 or 0	2/0 or 3/0	6	4
2/0 or 3/0	4/0 or 250 MCM	4	2
Over 3/0 thru 350 MCM	Over 250 MCM thru 500 MCM	2	0
Over 350 MCM thru 600 MCM	Over 500 MCM thru 900 MCM	0	3/0
Over 600 MCM thru 1100 MCM	Over 900 MCM thru 1750 MCM	2/0	4/0
Over 1100 MCM	Over 1750 MCM	3/0	250 MCM

Figure 11-14

12

Wiring Methods

Several types of wiring methods are currently in use for wiring installations. The methods used on a given project are determined by several factors:

1. The requirements set forth in the NE Code.
2. Local codes and ordinances.
3. Type of building construction.
4. Location of the wiring in the building.
5. Importance of the wiring system's appearance.
6. Costs and budget.

In general, two types of basic wiring methods are used in the majority of electrical systems: open and concealed wiring. In open wiring systems, the outlets and cable or raceway systems are installed on the surfaces of the walls, ceilings, columns, and the like where they are in view and readily accessible. Such wiring is often used in areas where appearance is not important and where it may be desirable to make frequent changes from time to time.

Concealed wiring systems have all cable and raceway runs concealed inside of walls, partitions, ceilings, columns, and behind baseboards or molding where they are out of view and not readily accessible. This type of wiring system is generally used in all new construction with finished interior walls, ceilings, and floors and is the preferred type where good appearance is important.

CABLE SYSTEMS

Several types of cable systems are used in electrical systems for building construction, and include the following:

Nonmetallic sheathed cable: Type NM cables are manufactured in two or three wires, and with varying sizes of conductors. The jacket or covering consists of rubber, plastic, or fiber. This type of cable may be concealed in the framework of buildings or, in some instances, may be run exposed on the building surfaces.

Underground feeder cable: Type UF cable may be used underground, including direct burial in the earth, as a feeder or branch-circuit cable when provided with overcurrent protection at the rated ampacity as required by the NE Code. When type UF cable is used above grade where it will come in direct contact with the rays of the sun, its outer covering must be sun resistant.

Service-entrance cable: Type SE cable, when used for an electrical service, must be installed as required by the NE Code, and may be used in interior wiring systems provided all the circuit conductors of the cable are insulated with rubber or thermoplastic insulation.

Armored cable: BX cable is manufactured in two-, three-, and four-wire assemblies, with varying sizes of conductors, and is used in locations similar to those where NM cable is used. The metallic spiral covering on BX cable offers a greater degree of mechanical protection than NM cable, and the metal jacket also provides for a continuously grounded system without the need for additional grounding conductors. This type of cable may be used for under-plaster extensions, as provided in the NE Code, and embedded in plaster finish, brick, or other masonry, except in damp or wet locations. It also may be run or fished in the air voids of masonry block or tile walls, except where such walls are exposed or subject to excessive moisture or dampness or are below grade.

Mineral-insulated metal-sheathed cable: Type MI cable is a factory assembly of one or more conductors insulated with a highly compressed refractory mineral insulation and enclosed in a liquid-tight and gas-tight continuous copper sheath. It may be used for services, feeders, and branch circuits in dry, wet, or continuously moist locations. Furthermore, it may be used indoors or outdoors, embedded in plaster, concrete, fill, or other masonry, whether above or below grade. This type of cable may also be used in hazardous locations, where exposed to oil or gasoline, where exposed to corrosive conditions not deteriorating to the cable's sheath, and in underground runs where suitably protected against physical damage and corrosive conditions. In other words, MI cable may be used in practically any electrical installation.

Power and control tray cable: Type TC power and control tray cable is a factory assembly of two or more insulated conductors, with or without associated bare or covered grounding conductors under a non-metallic sheath, approved for installation in cable trays, in raceways, or where supported by a messenger wire. The use of this cable is limited to industrial establishments where the conditions of maintenance and supervision assure that only qualified persons will service the installation.

Shielded nonmetallic cable: Type SNM cable is a factory assembly of two or more insulated conductors in an extruded core or moisture-resistant, flame-resistant nonmetallic material, covered with an over-lapping spiral metal tape and wire shield and jacketed with an extruded moisture-, flame-, oil-, corrosion-, fungus-, and sunlight-resistant non-metallic material. Type SNM cable may be used where operating temperatures do not exceed the rating marked on the cable, in cable trays or in raceways, and in hazardous locations where permitted in Articles 500 through 516 of the NE Code.

Metal-clad cable: Type MC cable is a factory assembly of one or more conductors, each individually insulated and enclosed in a metallic sheath of interlocking tape or a smooth or corrugated tube. This type of cable may be used for services, feeders, and branch circuits; power, lighting, control, and signal circuits; indoors or outdoors; where exposed or concealed; direct buried; in cable tray; in any approved raceway; as open runs of cable; as aerial cable on a messenger; in hazardous locations as permitted in Articles 501, 502, and 503 of the NE Code; in dry locations; and in wet locations under certain conditions as specified in the NE Code.

Flat conductor cable: Type FC cable is an assembly of parallel conductors formed integrally with an insulating material web specifically designed for field installation in metal surface raceway approved for the purpose. This type of cable may be used only as branch circuits to supply suitable tap devices for lighting, small appliances, or small power loads. Flat cable assemblies shall be installed for exposed work only. Flat cable assemblies shall be installed in locations where they will not be subjected to severe physical damage.

RACEWAY SYSTEMS

A raceway wiring system consists of an electrical wiring system in which two or more individual conductors are pulled into a conduit (pipe) or similar housing for the conductors after the raceway system has been completely installed. The basic raceways are rigid steel conduit, electrical metallic tubing (EMT), and PVC (polyvinyl chloride) plastic. Other raceways include surface metal moldings and flexible metallic conduit.

These raceways are available in standardized sizes and serve primarily to provide mechanical protection for the wires run inside and, in the case of metallic raceways, to provide a continuously grounded system. Metallic raceways, properly installed, provide the greatest degree of mechanical and grounding protection and provide maximum protection against fire hazards for the electrical system. However, they are more expensive to install.

Most electricians prefer to use a hacksaw with blade having 18 teeth per inch for cutting rigid conduit and 32 teeth per inch for cutting the smaller sizes of conduit. For cutting larger sizes of conduit (1-1/2 inches and above), a special conduit cutter should be used to save time. While quicker to use, the conduit cutter almost always leaves a hump inside the conduit and the burr is somewhat larger than made by a standard hacksaw. If a power band saw is available on the job, it is preferred for cutting the larger sizes of conduit. Abrasive cutters are also popular for the larger sizes of conduit.

Conduit cuts should be made square and the inside edge of the cut must be reamed to remove any burr or sharp edge that might damage wire insulation when the conductors are pulled inside the conduit. After reaming, most experienced electricians feel the inside of the cut with their finger to be sure that no burrs or sharp edges are present.

Lengths of conduit to be cut should be accurately measured for the size needed and an additional 3/8 inch should be allowed on the smaller sizes of conduit for terminations; the larger sizes of conduit will require approximately 1/2 inch for locknuts, bushings, and the like at terminations.

The usual practice for threading the smaller sizes of rigid conduit is to use a pipe vise and a die stock. The die stock should be fitted with proper-sized guides and sharp cutting dies properly adjusted and securely held in the stock. A good lubricant (cutting oil) is then used liberally during the cutting process. If sufficient lubricant is used, cuts may be made cleaner and sharper, and the cutting dies will last much longer.

Full threads must be cut to allow the conduit ends to come close together in the coupling or to firmly seat in the shoulders of threaded hubs of conduit bodies. To obtain a full thread, run the die up on the conduit until the conduit barely comes through the die. This will give a good thread length adequate for all purposes. Anything longer will not fit into the coupling and will later corrode because threading removes the zinc from the conduit.

Clean, sharply cut threads also make a better continuous ground and save much trouble once the system is in operation.

Electrical Metallic Tubing

Electrical metallic tubing (EMT) may be used for both exposed and concealed work except where it will be subjected to severe damage during use, in cinder concrete, or in fill where subjected to permanent moisture unless some means to protect it is provided; the tubing may be installed a minimum of 18 inches under the fill.

Threadless couplings and connectors are used for EMT installation and these should be installed so that the tubing will be made up tight. Where buried in masonry or installed in wet locations, couplings connectors, as well as supports, bolts, straps, and screws, should be of a type approved for the conditions.

Bends in the tubing should be made with a tubing bender so that no injury will occur and so the internal diameter of the tubing will not be effectively reduced. The bends between outlets or termination points should contain no more than the equivalent of four quarter-bends (360° total), including those bends located immediately at the outlet or fitting (offsets).

All cuts in EMT are made with either a hacksaw, power hacksaw, tubing cutter, or other approved device. Once out, the tubing ends should be reamed with a screwdriver handle or pipe reamer to remove all burrs and sharp edges that might damage conductor insulation.

Flexible Metal Conduit

Flexible metal conduit generally comes in two types, a standard metal-clad type and a liquid-tight type. The former type cannot be used in wet locations unless the conductors pulled in are of a type specially approved for such conditions. Neither type may be used where they will be subjected to physical damage or where any combination of ambient and/or conductor temperature will produce an operating temperature in excess of that for which the material is approved. Other uses are fully described in Articles 350 and 351 of the NE Code.

When this type of conduit is installed, it should be secured by an approved means at intervals not exceeding 4-1/2 feet and within 12 inches of every outlet box, fitting, or other termination points. In some cases, however, exceptions exist. For example, when flexible metal conduit must be finished in walls, ceilings, and the like, securing the conduit at these intervals would not be practical. Also, where more flexibility is required, lengths of not more than 3 feet may be utilized at termination points.

Flexible metal conduit may be used as a grounding means where both the conduit and the fittings are approved for the purpose. In lengths of more than 6 feet, it is best to install an extra grounding conductor within the conduit for added insurance.

Surface Metal Molding

When it is impractical to install the wiring in concealed areas, surface metal molding is a good compromise, even though it is visible, since proper painting to match the color of the ceiling and walls makes it very inconspicuous. Surface metal molding is made from sheet metal strips drawn into shape and comes in various shapes and sizes with factory fittings to meet nearly every application found around the home. A complete list of fittings can be obtained by writing The Wiremold Co., West Hartford, Connecticut 06110.

The running of straight lines of surface molding is simple. A length of molding with the coupling is slipped in the end, out enough so that the screw hole is exposed, and then the coupling is screwed to the surface to which the molding is to be attached. Then another length of molding is slipped on the coupling.

Factory fittings are used for corners and turns or the molding may be bent (to a certain extent) with a special bender. Matching outlet boxes for surface mounting are also available, and bushings are necessary at such boxes to prevent the sharp edges of the molding from injuring the insulation on the wire.

Clips are used to fasten the molding in place. The clip is secured by a screw and then the molding is slipped into the clip, wherever extra support of the molding is needed, and fastened by screws. When parallel runs of molding are installed, they may be secured in place by means of a multiple strap. The joints in runs of molding are covered by slipping a connection cover over the joints. Such runs of molding should be grounded the same as any other metal raceway, and this is done by use of grounding clips. The current-carrying wires are normally pulled in after the molding is in place.

The installation of surface metal molding requires no special tools unless bending the molding is necessary. The molding is fastened in place with screws, toggle bolts, and the like, depending on the materials to which it is fastened. All molding should be run straight and parallel with the room or building lines, that is, baseboards, trims, and other room moldings. The decor of the room should be considered first and the molding made as inconspicuous as possible.

It is often desirable to install surface molding not used for wires in order to complete a pattern set by other surface molding containing current-carrying wires, or to continue a run to make it appear to be part of the room's decoration.

WIREWAYS

Wireways are sheet-metal troughs with hinged or removable covers for housing and protecting wires and cables and in which conductors are held in place after the wireway has been installed as a complete system. They may be used only for exposed work and should not be installed where they will be subject to severe physical damage or corrosive vapor nor in any hazardous location except Class II, Division 2 of the NE Code.

The wireway structure must be designed to safely handle the sizes of conductors used in the system. Furthermore, the system should not contain more than 30 current-carrying conductors at any cross section. The sum of the cross-sectional areas of all contained conductors at any cross section of a wireway shall not exceed 20 percent of the interior cross-sectioned area of the wireway.

Splices and taps, made and insulated by approved methods, may be located within the wireway provided they are accessible. The conductors, including splices and taps, shall not fill the wireway to more than 75 percent of its area at that point.

Wireways shall be securely supported at intervals not exceeding 5 feet, unless specially approved for supports at greater intervals, but in no case shall the distance between supports exceed 10 feet.

BUSWAYS

There are several types of busways or duct systems for electrical transmission and feeder purposes. Lighting duct, trolley duct, and distribution bus duct are just a few. All are designed for a specific purpose, and the electrician or electrical designer should become familiar with all types before an installation is laid out.

Lighting duct, for example, permits the installation of an unlimited amount of footage from a single working platform. As each section and the lighting fixtures are secured in place, the complete assembly is then simply transported to the area of installation and installed in one piece.

Trolley duct is widely used for industrial applications, and where the installation requires a continuous polarization to prevent accidental reversal, polarizing bar is used. This system provides polarization for all trolley, permitting standard and detachable trolleys to be used on the same run.

Plug-in bus duct is also widely used for industrial applications, and the system consists of interconnected prefabricated sections of bus duct so formed that the complete assembly will be rigid in construction and neat and symmetrical in appearance.

CABLE TRAYS

Cable trays are used to support electrical conductors used mainly in industrial applications. The trays themselves are usually made up into a system of assembled, interconnected sections and associated fittings, all of which are made of metal or other noncombustible material. The finished system forms into a rigid structural run to contain and support single, multiconductor, or other wiring cables. Several styles of cable trays are available, including ladder, trough, channel, solid-bottom trays, and similar structures.

13

Service-Entrance Equipment

All buildings containing equipment that utilizes electricity require an electric service to enable the passage of electrical energy from the utility company's lines to the point of use within the buildings. Figure 13-1 shows the basic sections of an electric service. In this illustration, note that the high-voltage lines terminate on a power pole near the building that is being served. A transformer is mounted on the pole to reduce the voltage to a usable level (120/240 volts in this case). The remaining sections are described as follows:

> *Service drop:* The overhead conductors, through which electrical service is supplied, between the last power company pole and the point of their connection to the service facilities located at the building or other support used for the purpose.

> *Service entrance:* All components between the point of termination of the overhead service drop or underground service lateral and the building main disconnecting device, with the exception of the power company's metering equipment.

> *Service-entrance conductors:* The conductors between the point of termination of the overhead service drop or underground service lateral and the main disconnecting device in the building.

> *Service-entrance equipment:* Provides overcurrent protection to the feeder and service conductors, a means of disconnecting the feeders from energized service conductors, and a means of measuring the energy used by the use of metering equipment.

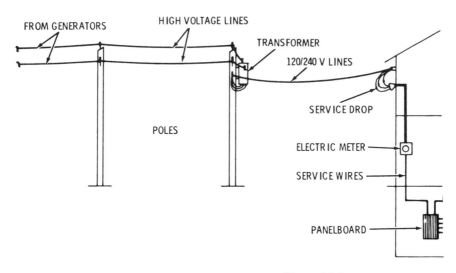

Figure 13-1

In cases where the service conductors to the building are routed underground, as shown in Fig. 13-2, these conductors are known as the service lateral, defined as follows:

> *Service lateral:* The underground conductors through which service is supplied between the power company's distribution facilities and the first point of their connection to the building or area service facilities located at the building or other support used for the purpose.

CLEARANCES FOR SERVICE DROPS AND LATERALS

The NE Code rules governing clearances for service drops and laterals are depicted in Fig. 13-3. In general, service-drop conductors must not be readily accessible and shall comply with the following NE Code requirements.

1. Conductors shall have a clearance of not less than 8 feet from the highest point of roofs over which they pass.

    Exception 1: Where the voltage between conductors does not exceed 300 and the roof has a slope of not less than 4 inches in 12 inches, a reduction in clearance to 3 feet shall be permitted.

    Exception 2: Where the voltage between conductors does not exceed 300, a reduction in clearance over the roof to not less than 18 inches shall be permitted if (1) they do not pass over more than 4 feet of the overhang portion of the roof, and (2) they are terminated at a through-the-roof raceway or approved support.

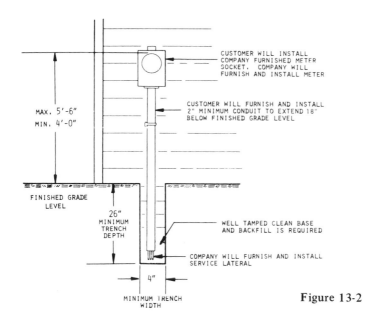

CUSTOMER WILL INSTALL
COMPANY FURNISHED METER
SOCKET. COMPANY WILL
FURNISH AND INSTALL METER

CUSTOMER WILL FURNISH AND INSTALL
2" MINIMUM CONDUIT TO EXTEND 18"
BELOW FINISHED GRADE LEVEL

MAX. 5'-6"

MIN. 4'-0"

FINISHED GRADE
LEVEL

26"
MINIMUM
TRENCH
DEPTH

WELL TAMPED CLEAN BASE
AND BACKFILL IS REQUIRED

COMPANY WILL FURNISH AND INSTALL
SERVICE LATERAL

4"

MINIMUM TRENCH
WIDTH

Figure 13-2

See Section 230-28 for mast supports.

2. Clearance from ground: Service-drop conductors when not in excess of 600 volts shall have the following minimum clearance from ground:
   - 10 feet above finished grade, sidewalks or from any platform or projection from which they might be reached.
   - 12 feet over residential driveways and commercial areas such as parking lots and drive-in establishments not subject to truck traffic.
   - 15 feet over commercial areas, parking lots, agricultural, or other areas subject to truck traffic.
   - 18 feet over public streets, alleys, roads, and driveways on other than residential property.
3. Clearance from building openings: Conductors shall have a clearance of not less than 3 feet from windows, doors, porches, fire escapes, or similar locations.

Conductors run above the top level of a window shall be considered out of reach from that window.

Note that the NE Code also provides rules for clearances for service drops from building openings such as windows and doors. Figure 13-4 summarizes these requirements.

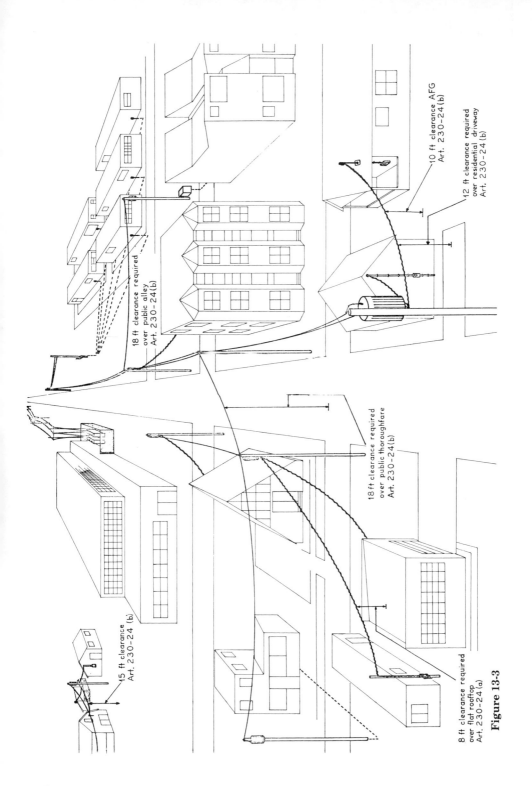

18 ft clearance required
over public alley
Art. 230-24(b)

10 ft clearance AFG
Art. 230-24(b)

12 ft clearance required
over residential driveway
Art. 230-24(b)

18 ft clearance required
over public thoroughfare
Art. 230-24(b)

15 ft clearance
Art. 230-24 (b)

8 ft clearance required
over flat rooftop
Art. 230-24 (a)

Figure 13-3

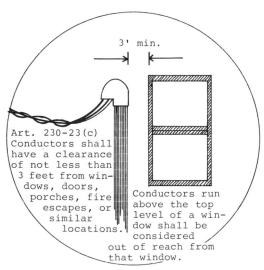

Figure 13-4

SERVICE CONNECTIONS

The point of attachment of conductors to a building or other structure must provide the minimum clearances as specified in Section 230-24 of the NE Code. In no case shall this point of attachment be less than 10 feet above finished grade.

Multiconductor cables used for service drops must be attached to buildings or other structures by fittings approved for the purpose. Open conductors shall be attached to fittings approved for the purpose or to noncombustible, nonabsorbent insulators securely attached to the building or other structure.

Figure 13-5 shows a drawing of a typical 200-ampere residential overhead service installation. To comply with NE Code requirements, the service is equipped with a raintight service head that is located above the point of attachment of the service-drop conductor to the building; the service cables are held securely in place by connection to the service drop conductors below the service head by a fitting approved for the purpose; the service-entrance cable is supported by straps within 12 inches of the service head and at intervals not exceeding 4-1/2 feet. Other requirements are shown in the drawings in Fig. 13-5.

Where a service mast is used for the support of the service conductors (see Fig. 13-6), it shall be of adequate strength or be supported by braces or guys to withstand safely the strain imposed by the service drop. The fittings must also be approved for the purpose.

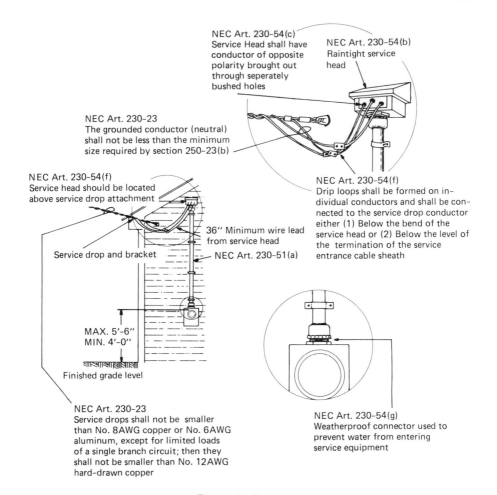

NEC Art. 230-54(c)
Service Head shall have conductor of opposite polarity brought out through seperately bushed holes

NEC Art. 230-54(b)
Raintight service head

NEC Art. 230-23
The grounded conductor (neutral) shall not be less than the minimum size required by section 250-23(b)

NEC Art. 230-54(f)
Service head should be located above service drop attachment

NEC Art. 230-54(f)
Drip loops shall be formed on individual conductors and shall be connected to the service drop conductor either (1) Below the bend of the service head or (2) Below the level of the termination of the service entrance cable sheath

36" Minimum wire lead from service head
NEC Art. 230-51(a)

Service drop and bracket

MAX. 5'-6"
MIN. 4'-0"

Finished grade level

NEC Art. 230-23
Service drops shall not be smaller than No. 8AWG copper or No. 6AWG aluminum, except for limited loads of a single branch circuit; then they shall not be smaller than No. 12AWG hard-drawn copper

NEC Art. 230-54(g)
Weatherproof connector used to prevent water from entering service equipment

Figure 13-5

SERVICE CONDUCTORS

Where the service voltages are below 600 volts, service-entrance conductors must be installed in accordance with the applicable requirements of the NE Code covering the type of wiring method used and limited to the following methods:

1. Open wiring on insulators (Fig. 13-7)
2. Rigid metal conduit (Fig. 13-8)
3. Intermediate metal conduit (Fig. 13-9)
4. Electrical metallic tubing (Fig. 13-10)

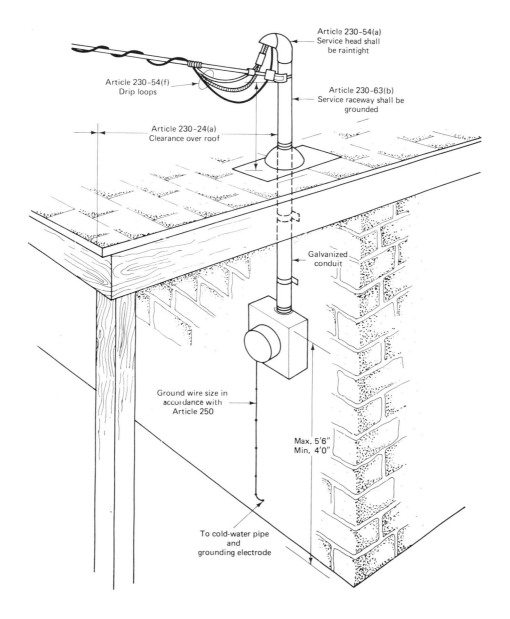

Article 230-54(a)
Service head shall
be raintight

Article 230-54(f)
Drip loops

Article 230-63(b)
Service raceway shall be
grounded

Article 230-24(a)
Clearance over roof

Galvanized
conduit

Ground wire size in
accordance with
Article 250

Max. 5'6"
Min. 4'0"

To cold-water pipe
and
grounding electrode

Figure 13-6

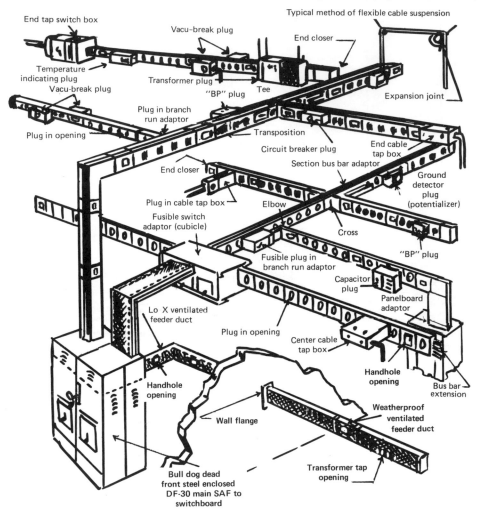

End tap switch box

Typical method of flexible cable suspension

Vacu-break plug

End closer

Temperature indicating plug

Transformer plug

Vacu-break plug

"BP" plug

Tee

Expansion joint

Plug in branch run adaptor

Transposition

End cable tap box

Plug in opening

Circuit breaker plug

Section bus bar adaptor

End closer

Ground detector plug (potentializer)

Plug in cable tap box

Elbow

Fusible switch adaptor (cubicle)

Cross

"BP" plug

Fusible plug in branch run adaptor

Capacitor plug

Panelboard adaptor

Lo X ventilated feeder duct

Plug in opening

Center cable tap box

Handhole opening

Handhole opening

Bus bar extension

Wall flange

Weatherproof ventilated feeder duct

Bull dog dead front steel enclosed DF-30 main SAF to switchboard

Transformer tap opening

ILLUSTRATION FOR BUSWAY

The perspective sketch of a typical BUStribution duct system shown above indicates numerous fittings available for use with standard duct systems. Also shown is the individual physical relationship between each fitting and rest of system.

Figure 13-7

5. Service-entrance cables (Fig. 13-11)
6. Wireways (Fig. 13-12)
7. Busways (Fig. 13-13)

Approved cable tray systems are permitted to support cables approved for use as service-entrance conductors. See Article 318.

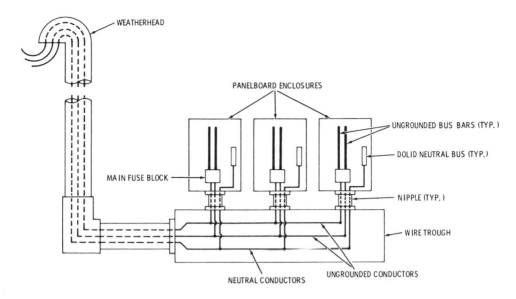

Figure 13-8

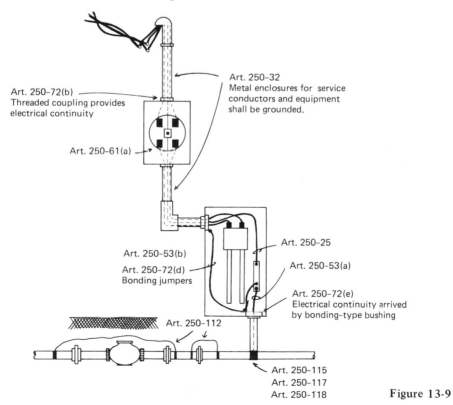

Figure 13-9

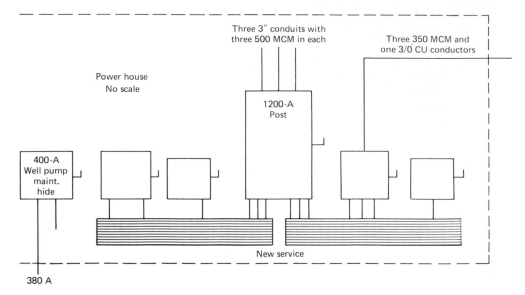

Power house
No scale

Three 3″ conduits with
three 500 MCM in each

Three 350 MCM and
one 3/0 CU conductors

1200-A
Post

400-A
Well pump
maint.
hide

New service

380 A

Figure 13-10

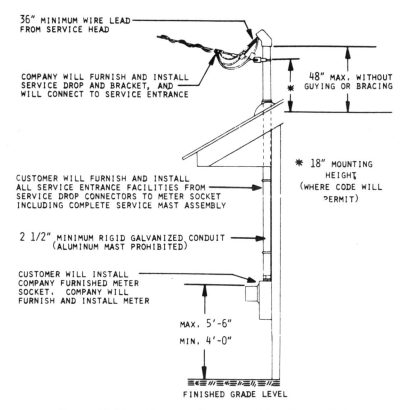

36″ MINIMUM WIRE LEAD
FROM SERVICE HEAD

COMPANY WILL FURNISH AND INSTALL
SERVICE DROP AND BRACKET, AND
WILL CONNECT TO SERVICE ENTRANCE

48″ MAX. WITHOUT
GUYING OR BRACING

✱ 18″ MOUNTING
HEIGHT
(WHERE CODE WILL
PERMIT)

CUSTOMER WILL FURNISH AND INSTALL
ALL SERVICE ENTRANCE FACILITIES FROM
SERVICE DROP CONNECTORS TO METER SOCKET
INCLUDING COMPLETE SERVICE MAST ASSEMBLY

2 1/2″ MINIMUM RIGID GALVANIZED CONDUIT
(ALUMINUM MAST PROHIBITED)

CUSTOMER WILL INSTALL
COMPANY FURNISHED METER
SOCKET. COMPANY WILL
FURNISH AND INSTALL METER

MAX. 5′-6″

MIN. 4′-0″

FINISHED GRADE LEVEL

Figure 13-11   (Courtesy Potomac Edison Power Co.)

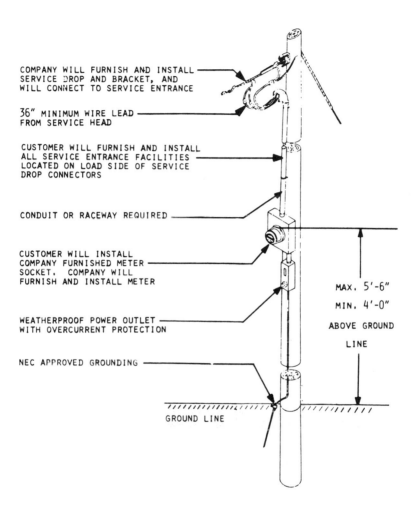

COMPANY WILL FURNISH AND INSTALL
SERVICE DROP AND BRACKET, AND
WILL CONNECT TO SERVICE ENTRANCE

36" MINIMUM WIRE LEAD
FROM SERVICE HEAD

CUSTOMER WILL FURNISH AND INSTALL
ALL SERVICE ENTRANCE FACILITIES
LOCATED ON LOAD SIDE OF SERVICE
DROP CONNECTORS

CONDUIT OR RACEWAY REQUIRED

CUSTOMER WILL INSTALL
COMPANY FURNISHED METER
SOCKET. COMPANY WILL
FURNISH AND INSTALL METER

WEATHERPROOF POWER OUTLET
WITH OVERCURRENT PROTECTION

NEC APPROVED GROUNDING

GROUND LINE

MAX. 5'-6"

MIN. 4'-0"

ABOVE GROUND

LINE

Figure 13-12   (Courtesy Potomac Edison Power Co.)

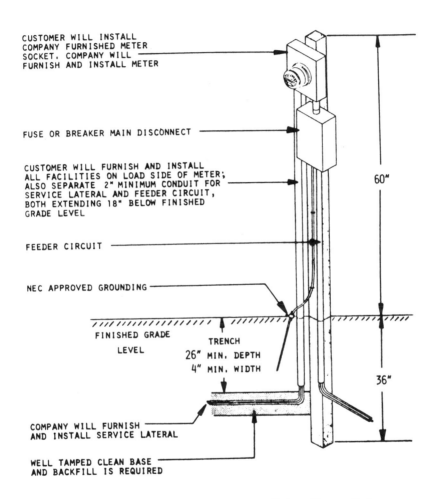

CUSTOMER WILL INSTALL
COMPANY FURNISHED METER
SOCKET. COMPANY WILL
FURNISH AND INSTALL METER

FUSE OR BREAKER MAIN DISCONNECT

CUSTOMER WILL FURNISH AND INSTALL
ALL FACILITIES ON LOAD SIDE OF METER;
ALSO SEPARATE 2" MINIMUM CONDUIT FOR
SERVICE LATERAL AND FEEDER CIRCUIT,
BOTH EXTENDING 18" BELOW FINISHED
GRADE LEVEL

FEEDER CIRCUIT

NEC APPROVED GROUNDING

FINISHED GRADE
LEVEL

TRENCH
26" MIN. DEPTH
4" MIN. WIDTH

COMPANY WILL FURNISH
AND INSTALL SERVICE LATERAL

WELL TAMPED CLEAN BASE
AND BACKFILL IS REQUIRED

60"

36"

Figure 13-13   (Courtesy Potomac Edison Power Co.)

GUARDING AND GROUNDING

Article 250 of the NE Code covers general requirements for grounding and bonding electrical services. Table 13-1 gives the proper sizes of grounding conductors for various sizes of electric services.

TABLE 13-1. Grounding Electrode Conductor for Grounded Systems

Size of Largest Service-Entrance Conductor or Equivalent for Parallel Conductors		Size of Grounding Electrode Conductor	
Copper	Aluminum or Copper-clad Aluminum	Copper	Aluminum or Copper-clad Aluminum
2 or smaller	0 or smaller	8	6
1 or 0	2/0 or 3/0	6	4
2/0 or 3/0	4/0 or 250 MCM	4	2
Over 3/0 through 350 MCM	Over 250 MCM through 500 MCM	2	0
Over 350 MCM through 600 MCM	Over 500 MCM through 900 MCM	0	3/0
Over 600 MCM through 1100 MCM	Over 900 MCM through 1750 MCM	2/0	4/0
Over 1100 MCM	Over 1750 MCM	3/0	250 MCM

In general, live parts of service equipment must be enclosed so that they will not be exposed to accidental contact or else guarded in accordance with Sections 110-17 and 110-18 of the NE Code, which read as follows:

110-17. Guarding of Live Parts.   (600 volts or less, nominal)
(a) Except as elsewhere required or permitted by this Code, live parts of electric equipment operating at 50 volts or more shall be guarded against accidental contact by approved cabinets or other forms of approved enclosures or by any of the following means:

(1) By location in a room, vault, or similar enclosure that is accessible only to qualified persons.

(2) By suitable permanent, substantial partitions or screens so arranged that only qualified persons will have access to the space within reach of the live parts. Any openings in such partitions or screens shall be so sized and located that persons are not likely to come into accidental contact with the live parts or to bring conducting objects into contact with them.

(3) By location on a suitable balcony, gallery, or platform so elevated and arranged as to exclude unqualified persons.

(4) By elevation of 8 feet or more above the floor or other working surface.
(b) In locations where electric equipment would be exposed to physical damage, enclosures or guards shall be so arranged and of such strength as to prevent such damage.
(c) Entrances to rooms and other guarded locations containing exposed live parts shall be marked with conspicuous warning signs forbidding unqualified persons to enter.

For motors, see Sections 430-132 and 430-133. For over 600 volts, see Section 110-34.

110-18. Arcing Parts.   Parts of electric equipment which in ordinary operation produce arcs, sparks, flames, or molten metal shall be enclosed or separated and isolated from all combustible material.

For hazardous locations, see Articles 500 through 517. For motors, see Section 430-14.

SERVICE EQUIPMENT

Service-entrance equipment is usually grouped at one centralized location. Feeders run to various locations to feed heavy-loaded electrical equipment and subpanels, which are located in a building to keep the length of the branch-circuit raceways at a practical minimum for operating efficiency and to cut down on cost.

The main service-disconnecting means will sometimes be made up on the job by assemblying individually enclosed fused switches on a length of metal auxiliary gutter. The various components are connected by means of short conduit nipples, in which the insulated conductors are installed. In other cases, the main disconnect and feeder overcurrent devices are enclosed in factory-assembled panelboard; the entire assembly is commonly called a main distribution panelboard.

Article 230-70 of the NE Code requires that a service be provided with a disconnecting means for all conductors in the building or structure from the service-entrance conductors. This disconnecting means should be located at or near the point where the service-entrance conductors enter the building. The NE Code also requires that all single feeders and branch circuits be provided with a means of individual disconnection from the source of supply. The disconnecting means must be located at a readily accessible point, either inside or outside the building, and adequate access and working space must be provided all around the disconnecting means.

Overcurrent protection is required both at the main source and for all individual feeders and branch circuits in order to protect the electrical installation against ground faults and overloads.

Figures 13-14 through 13-16 depict several types of service installations along with the applicable NE Code rules governing them.

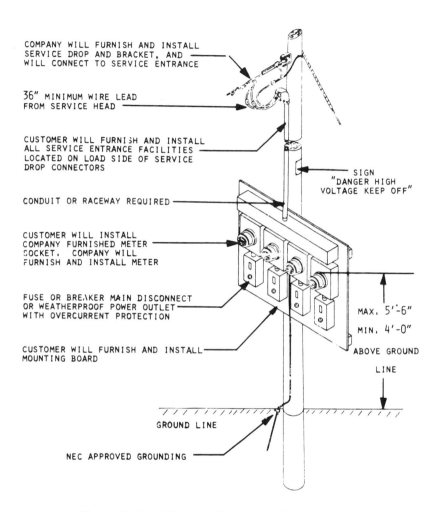

COMPANY WILL FURNISH AND INSTALL
SERVICE DROP AND BRACKET, AND
WILL CONNECT TO SERVICE ENTRANCE

36" MINIMUM WIRE LEAD
FROM SERVICE HEAD

CUSTOMER WILL FURNISH AND INSTALL
ALL SERVICE ENTRANCE FACILITIES
LOCATED ON LOAD SIDE OF SERVICE
DROP CONNECTORS

SIGN
"DANGER HIGH
VOLTAGE KEEP OFF"

CONDUIT OR RACEWAY REQUIRED

CUSTOMER WILL INSTALL
COMPANY FURNISHED METER
SOCKET. COMPANY WILL
FURNISH AND INSTALL METER

FUSE OR BREAKER MAIN DISCONNECT
OR WEATHERPROOF POWER OUTLET
WITH OVERCURRENT PROTECTION

MAX. 5'-6"
MIN. 4'-0"

ABOVE GROUND

CUSTOMER WILL FURNISH AND INSTALL
MOUNTING BOARD

LINE

GROUND LINE

NEC APPROVED GROUNDING

Figure 13-14   (Courtesy Potomac Edison Power Co.)

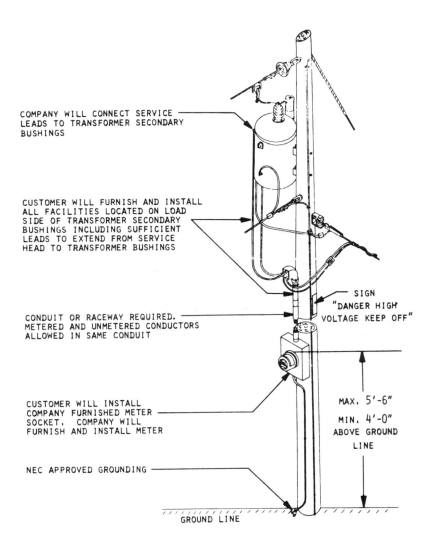

COMPANY WILL CONNECT SERVICE
LEADS TO TRANSFORMER SECONDARY
BUSHINGS

CUSTOMER WILL FURNISH AND INSTALL
ALL FACILITIES LOCATED ON LOAD
SIDE OF TRANSFORMER SECONDARY
BUSHINGS INCLUDING SUFFICIENT
LEADS TO EXTEND FROM SERVICE
HEAD TO TRANSFORMER BUSHINGS

CONDUIT OR RACEWAY REQUIRED.
METERED AND UNMETERED CONDUCTORS
ALLOWED IN SAME CONDUIT

CUSTOMER WILL INSTALL
COMPANY FURNISHED METER
SOCKET. COMPANY WILL
FURNISH AND INSTALL METER

NEC APPROVED GROUNDING

SIGN
"DANGER HIGH
VOLTAGE KEEP OFF"

MAX. 5'-6"

MIN. 4'-0"
ABOVE GROUND
LINE

GROUND LINE

Figure 13-15   (Courtesy Potomac Edison Power Co.)

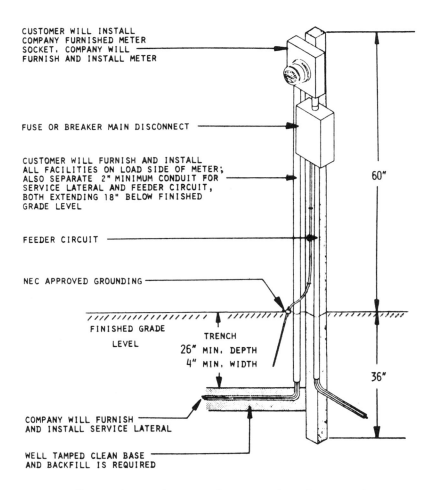

CUSTOMER WILL INSTALL
COMPANY FURNISHED METER
SOCKET. COMPANY WILL
FURNISH AND INSTALL METER

FUSE OR BREAKER MAIN DISCONNECT

CUSTOMER WILL FURNISH AND INSTALL
ALL FACILITIES ON LOAD SIDE OF METER;
ALSO SEPARATE 2" MINIMUM CONDUIT FOR
SERVICE LATERAL AND FEEDER CIRCUIT,
BOTH EXTENDING 18" BELOW FINISHED
GRADE LEVEL

FEEDER CIRCUIT

NEC APPROVED GROUNDING

60"

FINISHED GRADE
LEVEL

TRENCH
26" MIN. DEPTH
4" MIN. WIDTH

36"

COMPANY WILL FURNISH
AND INSTALL SERVICE LATERAL

WELL TAMPED CLEAN BASE
AND BACKFILL IS REQUIRED

Figure 13-16   (Courtesy Potomac Edison Power Co.)

SERVICE-ENTRANCE CALCULATIONS

The service-entrance conductors and equipment must have adequate ampacity in order to conduct safely the current for the loads supplied without causing a temperature to rise, which could damage the insulation on the conductors. The following methods describe how to calculate the ampacity for a number of different occupancies.

RESIDENCES

In general, a residence served with electric power can be supplied through only one set of service-entrance conductors (Article 230-2). The service conductors must have adequate mechanical strength and cannot be smaller than No. 8 AWG copper or No. 6 AWG aluminum (Article 230-23). The NE Code gives two methods for sizing service-entrance conductors for residences or dwellings: the standard method and the alternate method.

The standard method requires that 3 watts per square foot be utilized for the general lighting load. The measurement is taken from the outside building dimensions, and each two-wire, small-appliance circuit is rated at 1500 watts. After these first loads are calculated, a demand factor must be applied; that is, the first 3000 watts is rated at 100 percent, while the remaining watts of the general-lighting and small-appliance loads are rated at 35 percent.

The results of the calculation give the net computed load of the residence without electric range and major appliances. Therefore, all major appliances must be listed by their nameplate ratings, and the calculated load for the electric range must be given. All figures are totaled to find the total calculated load in watts. This figure is divided by the phase-to-phase voltage to determine the load in amperes so that the service-entrance conductors may be sized according to Article 310 of the NE Code.

Chapter 9 of the NE Code gives examples of how services for various types of residential occupancies are sized. To demonstrate the procedure, assume that a single-level residence measures 24 feet by 55 feet (outside dimensions) and requires four small-appliance circuits, an 8000-watt electric-range circuit, a 4500-watt electric water heater circuit, a 4500-watt electric clothes-dryer circuit, and a 5000-watt air-conditioner circuit.

The square footage must first be determined:

$$24 \times 55 = 1320 \text{ square feet}$$

Then establish the following table:

		Watts
General lighting load:		
1320 square feet at 3 watts		3,960
Appliance circuits:		
4 at 1500		6,000
	Total	9,960
Application of demand factor, Table 220-4(b) of the NE Code:		
3000 watts at 100 percent		3,000
9960 − 3000 watts = 6960 at 35 percent		2,436
Net computed load without range and major appliances		5,436
Electric range (see NE Code Table 220-5)		8,000
Water heater		4,500
Clothes dryer		4,500
Air conditioner		5,600
	Total calculated load	28,036

2. Amperes = 28,036/240 volts = 116.82
3. From tables in NE Code, Article 310, No. 2 AWG, THW copper wire will safely carry 115 amperes.
4. From tables in NE Code, Article 310, 1/0 aluminum THW conductors will carry 120 amperes.

In the optional method of calculating residential service-entrance conductor sizes, the general lighting and small-appliance loads are determined in the same manner as in the standard method. However, no diversity is taken until all the other loads have been determined. Furthermore, no diversity is allowed on the electric range: the nameplate rating must be figured in totaling the loads. All other major appliances are listed by their nameplate ratings and all the loads are totaled. The load calculation by the optional method of the residence used to demonstrate the standard method follows:

	Watts
General lighting load:	
1320 square feet at 3 watts	3,960
Small-appliance load:	
4 circuits at 1500	6,000
Electric range (nameplate rating)	12,000
Water heater	4,500
Clothes dryer	4,500
Air conditioner	5,600
Total connected load	36,560

The first 10,000 watts (10 kilowatts) must be rated at 100 percent (Article 220-7). The remaining load is calculated on the basis of a 40 percent diversity. Therefore,

	Watts
First 10 kW at 100 percent	10,000
Remaining load at 40 percent	
($36,560 - 10,000 = 26,560 \times 0.40$)	10,624
Total calculated load	20,624

To find the total load in amperes, divide the total calculated wattage by the phase-to-phase voltage.

$$\frac{20,624}{240} = 85.9 \text{ amperes}$$

Table 310-19 of the NE Code states that size No. 2 AWG aluminum conductors with THW insulation will safely carry the load. Footnote c of this table states that "for three-wire single-phase service the allowable ampacity of RH, RHH, RHW, and THW aluminum conductors is: No. 2, 100 amps.; No. 1, 110 amps." This is good because no modern residence should be supplied with less than 100-ampere service, regardless of the calculated load.

The NE Code states that the neutral conductor must be of sufficient size to carry the maximum unbalanced load determined by Section 220-4. The maximum unbalanced load is the maximum connected load between the neutral and any one ungrounded (phase) conductor. Therefore, no 240-volt two-wire circuit will need a neutral conductor, and loads for these circuits do not have to be included in sizing the neutral conductor. This permits the neutral conductor to be somewhat smaller than the phase conductors.

The neutral conductor for the residence used in the previous service-entrance calculations should be sized by the following procedure.

		Watts
General lighting and small-appliance load after applying demand factor (standard method)		9,960
Range load, 8000 watts at 70 percent (NE Code Table 220-11)		5,600
	Total	15,560

Then

$$\frac{15,560}{240} = 64.83 \text{ amperes}$$

Based on this calculation, the NE Code permits the installation of a neutral conductor capable of carrying 64.83 amperes as long as the wire size is not smaller than No. 4 copper or No. 6 aluminum. According to Table 310-79 of the NE Code, this requires a No. 4 AWG conductor with THW insulation.

PANELBOARDS AND SWITCHGEAR

Service switches or main distribution panelboards are normally installed at a point immediately where the service-entrance conductors enter the building. Branch circuit and feeder panelboards (when required in addition to the main service panelboard) are usually grouped together at one or more centralized locations to keep the length of the branch-circuit conductors at a practical minimum for operating efficiency and to lower the initial installation costs.

In electrical wiring installations for building construction, overcurrent protective devices, consisting of fuses or circuit breakers, are factory assembled in a metal cabinet, the entire assembly commonly being called a *panelboard.*

Sometimes, the main service-disconnecting means will be made up on the job by the workers by assembling individually enclosed fused switches or circuit breakers on a length of metal auxiliary gutter, as shown in Fig. 13-17. Note that the various components are connected by means of short conduit nipples in which the insulated conductors are fed. Other services will consist of one large panelboard, often called a main distribution panelboard, which gives a neater appearance.

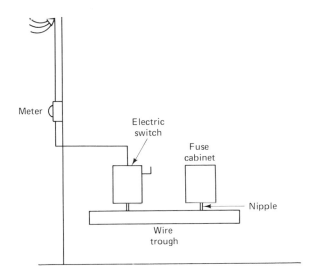

Figure 13-17

Panelboards consist of assemblies of overcurrent protective devices, with or without disconnecting devices, placed in a metal cabinet. The cabinet includes a cover or trim with one or two doors to allow access to the overcurrent and disconnecting devices and, in some types, access to the wiring space in the panelboard.

Panelboards fall into two mounting classifications: (1) flush mounting, wherein the trim extends beyond the outside edges of the cabinet to provide a neat finish with the wall surface, and (2) surface mounting, wherein the edge of the trim is flush with the edge of the cabinet.

Panelboards fall into two general classifications with regard to overcurrent protection devices: (1) circuit breaker and (2) fused. Small circuit breaker and fusible panelboards commonly referred to as *load centers* are manufactured for use in residential and small commercial and industrial occupancies.

OVERCURRENT PROTECTION

All electrical circuits and their related components are subject to destructive overcurrent. Harsh environments, general deterioration, accidental damage or damage from natural causes, excessive expansion, and overloading of the electrical system are factors that contribute to the occurrence of such overcurrents. Reliable protective devices prevent or minimize costly damage to transformers, conductors, motors, and the many other components and loads that make up the complete distribution system. Reliable circuit protection is essential to avoid the severe monetary losses that can result from power blackouts and prolonged downtime of facilities. To protect electrical conductors and equipment against abnormal operating conditions and their consequences, protective devices are used in circuits. The fuse and circuit breaker are two such devices.

Overcurrents

An overcurrent is either an overload current or a short-circuited current. The overload current is an excessive current relative to normal operating current, but one which is confined to the normal conductive paths provided by the conductors and other components and loads of the electrical system.

A short circuit (see Fig. 13-18) is probably the most common cause of electrical problems. It is an undesired current path that allows the electrical current to bypass the load on the circuit. Sometimes the short is between two wires due to faulty insulation, or it can occur between a wire and a grounded object, such as the metal frame of an electric range or clothes dryer.

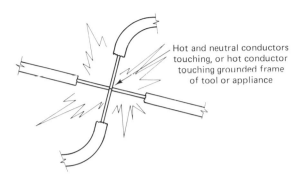

Hot and neutral conductors touching, or hot conductor touching grounded frame of tool or appliance

Figure 13-18

Overloads

Overloads are most often between one and six times the normal current level. Usually, they are caused by harmless temporary surge currents that occur when motors are started up or transformers are energized. Such overload currents normally occur. Since they are of brief duration, any temperature rise is trivial and has no harmful effect on the circuit components. (It is important that protective devices do not react to them.)

Continuous overloads can result from defective motors (such as worn motor bearings), overloaded equipment, or too many loads on one circuit. Such sustained overloads are destructive and must be cut off by protective devices before they damage the distribution system or affect system loads. However, since they are of relatively low magnitude compared to short-circuit currents, removal of the overload current within a few seconds will generally prevent equipment damage. A sustained overload current results in overheating of conductors and other components and will cause deterioration of insulation, which may eventually result in severe damage and short circuits if not interrupted.

FUSES

A fuse is the simplest device for opening an electric circuit when excessive current flows because of an overload or such fault conditions as grounds and short circuits. A fusible link or links encapsulated in a tube and connected to contact terminals comprise the fundamental elements of the basic fuse. Electrical resistance of the link is so low that it simply acts as a conductor, and every fuse is intended to be connected in series with each hot conductor so that current flowing through the conductor to any load must also pass through the fuse. The continuous current rating of the fuse in amperes establishes the maximum amount of current the fuse will carry without opening. When circuit current flow exceeds this value, an internal element (link) in the fuse melts due to the heat of the current flow and opens the circuit.

   Fuses are made in a wide variety of types and sizes with different current ratings, different abilities to interrupt fault currents, various speeds of operation (either quick opening or time-delay opening), different internal and external construction, and voltage ratings for both low-voltage (600 volts and below) and medium-voltage (over 600 volts) circuits.

CIRCUIT BREAKERS

A circuit breaker resembles an ordinary toggle switch, and it is probably the most widely used means of overcurrent protection today. On an overload, the circuit breaker opens itself or *trips.* In a tripped position, the handle jumps to the middle position (Fig. 13-19). To reset, turn the handle to the OFF position and then turn it as far as it will go beyond this position; finally, turn it to the ON position.

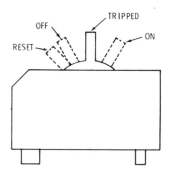

Figure 13-19

   On a conventional 120/240 volt, single-phase service, one single-pole breaker protects a 120-volt circuit, and one double-pole breaker protects a 240-volt circuit. The breakers are rated in amperes, just like fuses, although the particular ratings are not exactly the same as those for fuses.
   Circuit-breaker enclosures come in several types; one contains only branch-circuit breakers, while another contains a main-circuit breaker in addition to branch-circuit breakers. Most of the types used for residential applications are of the plug-in type; that is, the cabinets are usually sold without the breakers but contain an arrangement of bus bars. The user then selects the combination of breakers required and plugs them into this bus-bar arrangement.
   If the circuit-breaker cabinet contains six or less circuit breakers, it is permissible to eliminate a main-disconnecting means, provided the breakers are rated at more than 20 amperes. However, most of the circuits used in residential wiring are rated at 15 or 20 amperes, which means that a main circuit breaker would be required.

14

Principles of Illumination

Lighting fixtures or luminaires provide a means for supporting light sources (lamps) and are also used to control the light given off by the lamps. Lighting fixtures range from a simple porcelain lampholder (Fig. 14-1) to highly sophisticated chandeliers, as shown in Fig. 14-2.

Light sources include incandescent filament lamps, fluorescent lamps, high-intensity discharge lamps, short-arc lamps, miscellaneous discharge lamps (such as low-pressure sodium lamps), electroluminescent, light-emitting diodes, and nuclear light sources.

Lighting fixtures may be installed as individual units or grouped to form continuous rows or other patterns. The fixtures may be surface mounted (Fig. 14-3), suspended from the ceiling (Fig. 14-4), or recessed above the surface of the ceiling, as shown in Fig. 14-5. Another popular application is to install the lighting fixtures behind horizontal or vertical coves, which conceal the fixtures and provide indirect lighting.

For general lighting, lighting fixtures are usually located symmetrically in each bay or pair of bays, although this practice is often modified by the position of the existing outlets (in the case of modernization projects), the general architecture of the area, the type of luminaire, or the nature and location of the seeing tasks to be performed. Where working positions are fixed, it is sometimes desirable to position the luminaires with reference to the particular areas where high intensities are necessary. In schools, offices, or other areas where work is performed throughout the room, a high degree of uniformity is desirable. In such cases the spacing from the wall is normally one-half the distance between luminaires, as shown in the floor plan in Fig. 14-6. If desks or work benches are next to the walls, one-third the luminaire spacing is customary.

Figure 14-1   Typical porcelain lampholder, the least expensive lighting fixture that can be installed.

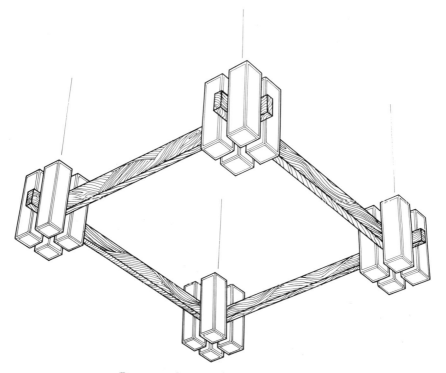

Figure 14-2   An elaborate chandelier.

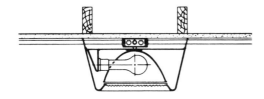

Installation over
surface outlet box

Option "−69" provides K.O.'s
in fixture ceiling pan for exposed 1/2" and 3/4" conduit.

Figure 14-3   Surface-mounted lighting fixture.

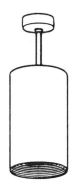

Figure 14-4   Pendant-mounted lighting fixture.

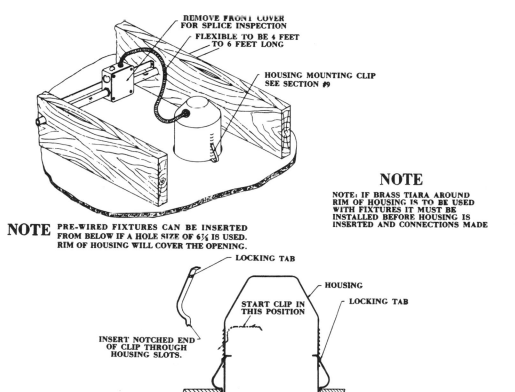

REMOVE FRONT COVER
FOR SPLICE INSPECTION

FLEXIBLE TO BE 4 FEET
TO 6 FEET LONG

HOUSING MOUNTING CLIP
SEE SECTION #9

NOTE PRE-WIRED FIXTURES CAN BE INSERTED
FROM BELOW IF A HOLE SIZE OF 6⅜ IS USED.
RIM OF HOUSING WILL COVER THE OPENING.

NOTE

NOTE: IF BRASS TIARA AROUND
RIM OF HOUSING IS TO BE USED
WITH FIXTURES IT MUST BE
INSTALLED BEFORE HOUSING IS
INSERTED AND CONNECTIONS MADE

LOCKING TAB

HOUSING

LOCKING TAB

START CLIP IN
THIS POSITION

INSERT NOTCHED END
OF CLIP THROUGH
HOUSING SLOTS.

CUT HOLE IN CEILING
APPROX. 6¼" (TEMPLET IN CARTON)

Figure 14-5   Recessed lighting fixture.

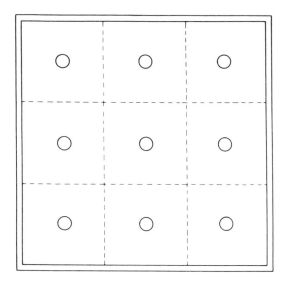

Figure 14-6   Method of correctly spacing lighting fixtures. Note that the distance from the wall to the fixtures is half the distance of that between the lighting fixtures.

LIGHTING CALCULATIONS

The quantity of light for an area may be calculated in many ways. Article 220 of the NE Code provides requirements for determining the number of branch circuits required for lighting loads and also for computing lighting loads.

    The number of circuits needed for general lighting in various occupancies is based on a unit-load requirement of not less than that specified in Table 220-2(b) of the NE Code (see Fig. 14-7). This table lists the minimum lighting load for each square foot of floor area for a particular occupancy. The floor area for each level must be computed from the outside dimensions of the building or area involved. For dwelling units, the computed floor area does not have to include open porches, garages, or unused or unfinished spaces not adaptable for future use.

    To illustrate the use of Table 220-2(b) of the NE Code, refer to the drawing in Fig. 14-8. This is a store building with outside dimensions of 36 by 72 feet, or 2592 square feet. Table 220-2(b) gives a unit-load requirement of 3 watts per square foot for stores, so the calculations are as follows:

$$2592 \times 3 = 7776 \text{ watts}$$

    However, since the store lighting will be in operation for over 3 hours at a time, the lighting load in this case is considered by the NE Code to be a continuous load. Therefore, the loads on the branch circuits feeding these

Type of Occupancy	Unit Load per Sq. Ft. (Watts)
Armories and Auditoriums	1
Banks	5
Barber Shops and Beauty Parlors	3
Churches	1
Clubs	2
Court Rooms	2
*Dwelling Units	3
Garages — Commercial (storage)	½
Hospitals	2
*Hotels and Motels, including apartment houses without provisions for cooking by tenants	2
Industrial Commercial (Loft) Buildings	2
Lodge Rooms	1½
Office Buildings	5
Restaurants	2
Schools	3
Stores	3
Warehouses (storage)	¼

In any of the above occupancies except one-family dwellings and individual dwelling units of multifamily dwellings:
    Assembly Halls and Auditoriums
    Halls, Corridors, Closets
    Storage Spaces
	
	1
	½
	¼

Figure 14-7   NE Code lighting requirements for various occupancies.

lighting fixtures must not exceed 80 percent of the branch-circuit rating. To modify the initial wattage, we multiply the initial answer by 1.20. Thus,

$$7776 \times 1.20 = 9331 \text{ watts}$$

$$\frac{\text{Watts}}{\text{Volts}} = \frac{9331}{120} = 77.76 \text{ amperes}$$

This indicates that four 20-ampere branch circuits will suffice, according to the NE Code.

Note that this calculation is based on minimum load conditions and 100 percent power factor, and may not provide sufficient capacity for the installation contemplated. For example, Fig. 14-9 shows the lighting layout for the store area in Fig. 14-8 as laid out by an electrical designer. Forty

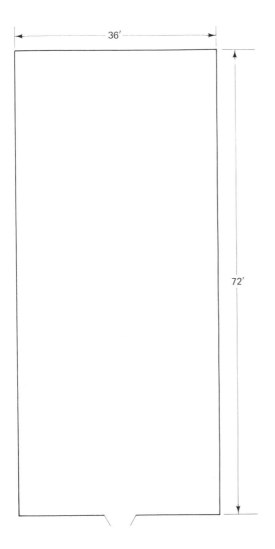

36′

72′

Figure 14-8   Floor plan of a typical commercial building.

2- by 4-foot lighting fixtures, each containing four 40-watt fluorescent lamps, are used in the store area. This constitutes a total load of approximately 8000 watts (lamps rated at 6400 watts plus an additional 1200 watts for the ballasts). To compensate for a continuous load,

$$8000 \times 1.20 = 9600 \text{ watts}$$

Four 20-ampere circuits would have still sufficed for this design, but the designer chose to use six 20-ampere circuits so as not to load each circuit to its maximum.

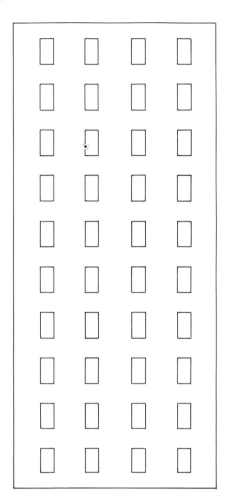

Figure 14-9   Commercial building in Fig. 14-8 after lighting fixtures have been laid out.

Article 210-22 of the NE Code states: For circuits supplying lighting units having ballasts, transformers, or autotransformers, the computed load shall be based on the total ampere ratings of such units and not on the total watts of the lamp.

The store building just described illustrates the application of this NE Code Article. For example, the total wattage of the lamps used in the store building is 6400 watts. However, an allowance must be made for the ballast. A good rule of thumb is to allow 25 percent additional wattage for ballast. Therefore, 6400 watts × 1.25 = 8000 watts.

Like most stores, the one under consideration has a show window for displaying its merchandise. For show-window lighting, a load of not less than 200 watts shall be included for each linear foot of show window, measured horizontally along its base (NE Code Article 220-12).

This show window has 13 linear feet and will therefore require a circuit capable of handling 2600 watts $\times$ 1.20 (allowance for continuous load) = 3120 watts.

$$\frac{\text{Watts}}{\text{Volts}} = \frac{3120}{120} = 26 \text{ amperes}$$

Therefore, the show window will require at least two 20-ampere branch circuits for the lighting as required by Article 220-12 of the NE Code.

The fact that 3120 watts has been calculated as the required load for this particular show-window lighting does not mean that this amount of power will be used. However, two circuits must be provided and the 3120-watt figure must be used in calculating the size of the service and feeders for this store building.

In all cases, the actual amount of power used by the show-window lights will be the wattage of the individual lighting fixture multiplied by the total number of fixtures. If the total wattage of the fixtures actually used exceeds the calculated amount, Article 210-22 states: "The total load shall not exceed the rating of the branch circuit. . . ." Therefore, more branch circuits may have to be used in some cases.

Table 220-11 Lighting Load Feeder Demand Factors		
Type of occupancy	Portion of Lighting Load to Which Demand Factor Applies (Wattage)	Demand Factor Percent
Dwelling units	First 3000 or less at	100
	Next 3001 to 120,000 at	35
	Remainder over 120,000 at	25
Hospitals*	First 50,000 or less at	40
	Remainder over 50,000 at	20
Hotels and motels, including apartment houses without provision for cooking by tenants*	First 20,000 or less at	50
	Next 20,001 to 100,000 at	40
	Remainder over 100,000 at	30
Warehouses (storage)	First 12,500 or less at	100
	Remainder over 12,500 at	50
All others	Total wattage	100

*The demand factors of this table shall not apply to the computed load of feeders to areas in hospitals, hotels, and motels where the entire lighting is likely to be used at one time, as in operating rooms, ballrooms, or dining rooms.

Figure 14-10 Reprinted with permission from NFPA 70-1984, Copyright ©1983, National Electrical Code®, National Fire Protection Association, Quincy, MA 02269. This reprinted material is not the complete and official position of the NFPA on the referenced subject, which is represented only by the standard in its entirety. National Electrical Code® and NEC® are Registered Trademarks of the National Fire Protection Association, Inc., Quincy, MA.

The demand factors listed in the table in Fig. 14-10 may be used for certain occupancies in calculating feeder sizes for lighting panelboards. Note, however, that the lighting load for the store building under consideration must be rated at 100 percent. Only dwelling units, hospitals, hotels, motels, and warehouses are allowed a demand factor under the present edition of the NE Code.

To illustrate the application of demand factors to a dwelling unit, assume that a residence has a total living area (excluding garages and the like) of 1500 square feet. Since the unit load specified in the NE Code Table 220-2(b) is 3 watts per square foot, the total lighting requirement equals 4500 watts. The feeder circuit for this panel may then be sized according to NE Code Table 220-11 as follows:

> Feeder demand factor (NE Code Table 220-11)
> First 3000 watts at 100%   3000 watts
> Remaining 1500 watts at 35%   525 watts
>   Total watts   3525 watts

Of course, in an actual application, other circuits, like small-appliance circuits, will enter into the calculation, but the principle is the same; that is, total all lighting, appliance, and similar circuits, and then use Table 220-11 to apply the demand factor, giving the total demand load in watts.

RESIDENTIAL LIGHTING

When designing or installing a lighting system for a dwelling unit, first calculate the lighting load as discussed previously. Lighting outlets are then located and lighting fixtures are selected to provide the highest visual comfort and performance that is consistent with the type of area in the home to be illuminated. The budget provided and the architectural and decorating schemes also play important roles in the design of a residential lighting layout.

Figure 14-11 shows a floor plan of a residential bedroom. Article 410-8 of the NE Code covers the requirements for the use of lighting fixtures in clothes closets. Figure 14-12 summarizes the following requirements:

(a) Location. A fixture in a clothes closet shall be permitted to be installed:
(1) On the wall above the closet door, provided the clearance between the fixture and a storage area where combustible material may be stored within the closet is not less than 18 inches, or
(2) On the ceiling over an area which is unobstructed to the floor, maintaining an 18-inch clearance horizontally between the fixture and a storage area where combustible material may be stored within the closet.
  A flush recessed fixture equipped with a solid lens shall be considered outside the closet area. [See Fig. 14-13.]
(b) Pendants. Pendants [Fig. 14-14] shall not be installed in clothes closets.

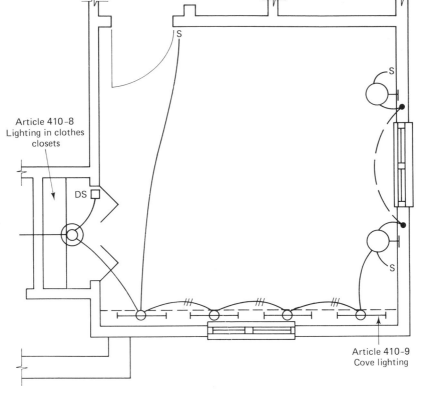

Figure 14-11   Floor plan of a residential bedroom showing location of lighting fixtures and related circuits.

The cove lighting in Fig. 14-11 contains four fluorescent fixtures. Article 410-9 states that coves shall have adequate space and shall be so located that lamps and equipment can be properly installed and maintained (see Fig. 15-15).

The fluorescent fixtures used in the cove are obviously surface mounted and each contains a ballast. If these fixtures were to be installed on combustible low-density cellulose fiberboard, Article 410-76(b) states: "it [the fixture] shall be approved for this condition or shall be spaced not less than 1-1/2 inches from the surface of the fiberboard. Where such fixtures are partially or wholly recessed, the provisions of Sections 410-64 through 410-72 shall apply."

The floor plan of the residence in Fig. 14-16 shows three lighting fixtures installed outside the building, two wall bracket fixtures at the front door, and one surface-mounted fixture on the carport ceiling. Article 410-4(a) states:

> Fixtures installed in wet or damp locations shall be approved for the purpose and shall be so constructed or installed that water cannot enter or accumulate in

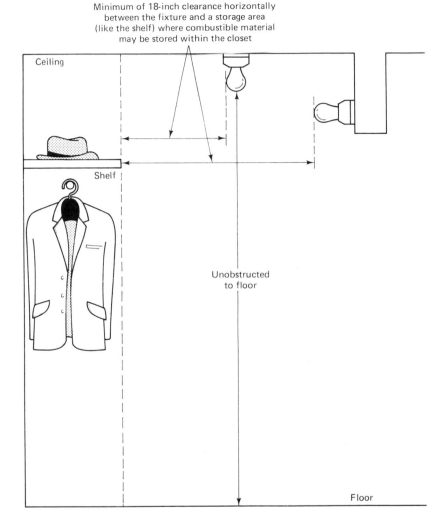

Minimum of 18-inch clearance horizontally
between the fixture and a storage area
(like the shelf) where combustible material
may be stored within the closet

Ceiling

Shelf

Unobstructed
to floor

Floor

Figure 14-12 Summary of NE Code requirements for residential
clothes closets.

wireways, lampholders, or other electrical parts. All fixtures installed in wet locations
shall be marked, "Suitable for Wet Locations." All fixtures installed in damp locations
shall be marked, "Suitable for Wet Locations" or "Suitable for Damp Locations."

Interior locations protected from weather but subject to moderate degrees of
moisture, such as some basements, some barns, some cold storage warehouses and
the like, the partially protected locations under canopies, marquees, roofed open
porches, and the like, shall be considered to be damp locations with respect to the
above requirement.

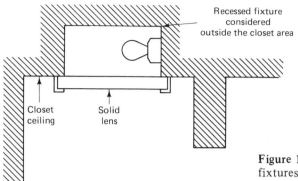

Figure 14-13 Recessed lighting fixtures are considered to be outside the closed area.

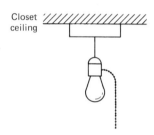

Figure 14-14 Pendant-mounted lighting fixtures cannot be used in residential clothes closets.

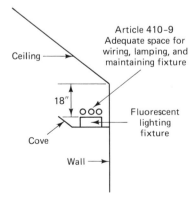

Figure 14-15 NE Code requirement for fluorescent cove lighting.

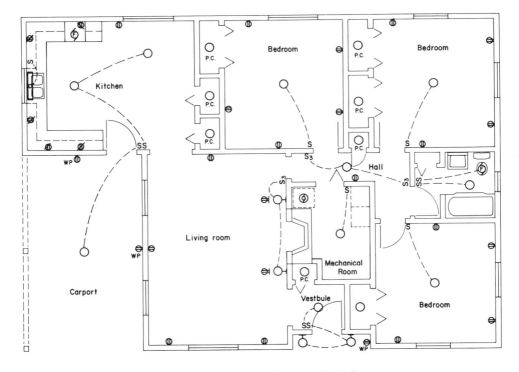

Figure 14-16   Lighting layout for a residential occupancy.

Article 410-16
Lighting fixture weighing
over 50 pounds must be
supported independently
of the outlet box

Figure 14-17   Lighting fixtures weighing over 50 pounds must be supported independently of the outlet box. (Courtesy Hadco Lighting.)

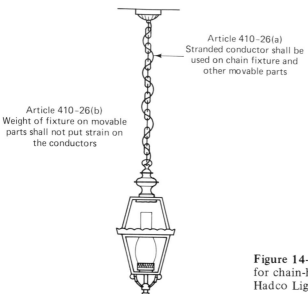

Article 410–26(a)
Stranded conductor shall be
used on chain fixture and
other movable parts

Article 410–26(b)
Weight of fixture on movable
parts shall not put strain on
the conductors

Figure 14-18   NE Code requirement for chain-hung fixtures. (Courtesy Hadco Lighting.)

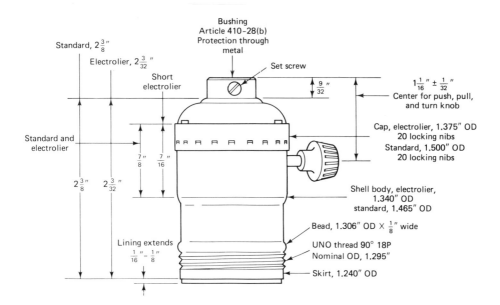

Bushing
Article 410–28(b)
Protection through
metal

Standard, $2\frac{3}{8}''$

Electrolier, $2\frac{3}{32}''$

Short
electrolier

Set screw

$\frac{9}{32}''$

$1\frac{1}{16}'' \pm \frac{1}{32}''$
Center for push, pull,
and turn knob

Standard and
electrolier

Cap, electrolier, 1.375" OD
20 locking nibs
Standard, 1.500" OD
20 locking nibs

$\frac{7}{8}''$ $\frac{7}{16}''$

$2\frac{3}{8}''$ $2\frac{3}{32}''$

Shell body, electrolier,
1.340" OD
standard, 1.465" OD

Lining extends
$\frac{1}{16}'' - \frac{1}{8}''$

Bead, 1.306" OD × $\frac{1}{8}''$ wide

UNO thread 90° 18P
Nominal OD, 1.295"

Skirt, 1.240" OD

Figure 14-19   NE Code requirements for metal lampholders. (Courtesy Leviton.)

Most lighting fixtures used in residential construction are mounted directly to the outlet box. However, in some cases, a fixture weighing more than 50 pounds, like the one in Fig. 14-17, will be used. In this case, Article 410-16 requires that the fixture be supported independently of the outlet box.

Lighting fixtures hung on fixture chains (as shown in Fig. 14-18) or other movable or flexible parts must have the fixture wires arranged so that the weight of the fixture or movable part shall not put a strain on the conductors (Article 410-26).

Metal lampholders, such as the one in Fig. 14-19, must provide means of securing the conductors in a manner that will not tend to cut or abrade the insulation [Article 410-28(a)]. Furthermore, where a metal lampholder is attached to a flexible cord, the inlet shall be equipped with an insulating bushing, which, if threaded, shall not be smaller than nominal 3/8-inch pipe size. The cord hole shall be of a size appropriate for the cord, and all burrs and fins shall be removed in order to provide a smooth bearing surface for the cord.

Bushings having holes 9/32 inch in diameter shall be permitted for use with plain pendant cord and holes 13/32 inch in diameter with reinforced cord.

RESIDENTIAL LIGHTING CALCULATIONS

Residential lighting design does not require the elaborate calculations that might be required for a commercial or industrial application. However, the designer must apply talent and ingenuity in selecting the best types of lighting fixtures for the various locations. He or she must achieve not only the proper quantity of illumination but also a desirable effect that will complement the architectural style of the house.

The following definitions should be learned in order to better understand the method of residential lighting design:

Lumen: A lumen is the quantity of light that will strike a surface of one square foot, all points of which are a distance of one foot from a light source of one candlepower.

Footcandle: A footcandle is a unit of measurement that represents the intensity of illumination that will be produced on a surface that is one foot distant from a source of one candlepower, and at right angles to the light rays from the source.

From these statements, we can say that one lumen per square foot equals one footcandle. Thus, this method will be called the lumens-per-square-foot method of residential lighting design.

In using this method, it is important to remember that lighter room colors reflect light and darker colors absorb light. This method is based on rooms with light colors; therefore, if the room surfaces are dark, the total lumens should be multiplied by a factor of 1.25.

Table 14-1 gives the required lumens per square foot for various areas in the home and also the required illumination in footcandles for those who desire to use a different method of calculating required illumination. The recommended footcandle level is fixed and will apply regardless of the type of lighting fixtures used. However, the recommended lumens given in this table are based on the assumption that portable table lamps, surface-mounted fixtures, or efficient structural lighting techniques will be used. If the majority of the lighting fixtures in an area will be recessed, the lumen figures in the table should be multiplied by 1.8. Note that this method produces only approximate results.

TABLE 14-1.

Area	Lumens Required (per ft^2)	Average Footcandles Required
Living room	80	70
Dining room	45	30
Kitchen	80	70
Bathroom	65	50
Bedroom	70	30
Hallway	45	30
Laundry	70	50
Workbench	70	70

Example 1:

Determine the total lumens required to achieve 80 lumens per square foot in the living room of the residence.

Solution.   The first step is to scale the drawings to find the dimensions of the area in question. In doing so, we find that the area is 13.75 feet wide and 19 feet long. Thus, 13.75 × 19 = 261.25, which is rounded off to 261 square feet. The preceding table recommends 80 lumens per square foot. Therefore, 80 × 261 = 20,880 lumens required in this area.

The next step is to refer to a lamp catalog in order to select lamps that will give the required total lumens. At the same time, the designer should have a good idea of the type of lighting fixtures that will be used, as well as their location.

In the floor plan of the residence (Fig. 14-20), we decided on two recessed spotlights mounted in the ceiling above the fireplace. Each fixture will contain two 75-watt R-30 lamps at 860 lumens each, for a total of 1720 lumens. However, since these are recessed, the total lumens will have to be multiplied by 0.555 to obtain the effective lumens.

$$1720 \times 0.555 = 955 \text{ effective lumens}$$

This means that we now need 19,925 more lumens in this area in order to meet the recommended level of illumination.

The next section of this area will be the window area on the front side of the house. It was decided to use a drapery cornice from wall to wall, which would contain four 40-watt warm-white fluorescent lamps at 2080 lumens each, for a total of 8320 lumens. Combining this figure with the 955 effective lumens from the recessed lamps gives a total of 9275 lumens, or 11,605 more lumens to account for.

Two 3-way (100-, 200-, 300-watt) bulbs in table lamps will be used on end tables, one on each side of the sofa for a total of 9460 lumens; this means that only 2145 lumens are unaccounted for. However, one 3-way (50-, 100-, 150-watt) bulb will be used in a lamp on a chairside table. Since the lamp is rated at 2190 lumens, we now have the total lumens required for the area.

It can be seen that this method of residential lighting calculation makes it possible to quickly determine the light sources needed to achieve the recommended illumination level in any area of the home.

The living room is the room where guests are entertained and where the family gathers to relax or engage in conversation. The living-room lighting should emphasize any special architectural features such as fireplaces or bookcases. This also holds true for special room accents, such as draperied walls, planters, and paintings.

Dramatizing fireplaces with accent lights brings out the texture of bricks, adds to overall room light level, and eliminates bright spots that cause subconscious irritation over a period of time. Recessed lighting fixtures using from 75- to 150-watt lamps are excellent for this application.

A row of recessed downlights or cornice or valance lighting all add life to draperies, paneled walls, and the like. They also supplement the general room-lighting level with glare-free light.

Downlights used for living-room accent lighting should be spaced approximately 2-1/2 to 3 feet apart and 8 to 10 inches from the wall. Valances are nearly always used at windows or above draperied walls. They provide uplight, which reflects off the ceiling for general room lighting, and downlight for drapery accent. Cornices, on the other hand, direct all of their light downward to give dramatic interest to wall coverings, draperies, and so on, and are also good for low-ceiling rooms.

Pull-down fixtures or table lamps are used for reading areas. While pull-down fixtures are more dramatic, the designer must know the furniture placement prior to locating the fixtures. The pull-down fixtures then have the disadvantage of not permitting furniture movement under these areas at a later time.

As a final touch to living-room lighting, dimmers should be added to vary the lighting levels exactly to the living-room activities—low for a relaxed mood, bright for a happy party mood.

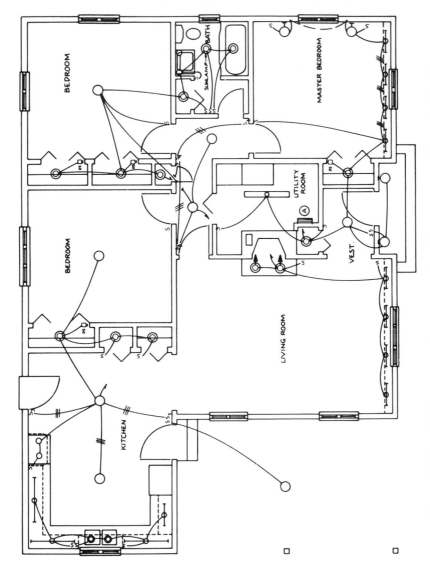

Figure 14-20 Lighting layout of a typical residential occupancy. (From *Residential Electrical Design*, courtesy Craftsman Book Co.)

ZONAL-CAVITY METHOD

The zonal-cavity method of calculating average illumination levels assumes each room or area to consist of the following three separate cavities: ceiling cavity, room cavity, and floor cavity.

Figure 14-21 shows that the ceiling cavity extends from the lighting fixture plane upward to the ceiling. The floor cavity extends from the work plane downward to the floor, while the room cavity is the space between the lighting fixture plane and the working plane.

If the lighting fixtures are recessed or surface-mounted on the ceiling, there will be no ceiling cavity and the ceiling-cavity reflectance will be equal to the actual ceiling reflectance. Similarly, if the work plane is at floor level, there will be no floor cavity and the floor-cavity reflectance will be equal to the actual floor reflectance. The geometric proportions of these spaces become the *cavity ratios*.

Cavity Ratio

Rooms are classified according to shape by ten cavity-ratio numbers. The basic formula for obtaining cavity ratios in rectangular-shaped rooms is

$$\text{Cavity ratio} = \frac{5 \times \text{height (length + width)}}{\text{length} \times \text{width}}$$

where height is the height of the cavity under consideration, that is, ceiling, floor, or room cavity.

For example, assume the room illustrated in Fig. 14-21 is 8 feet wide by 12 feet in length. The lighting fixtures are suspended 1 foot below the

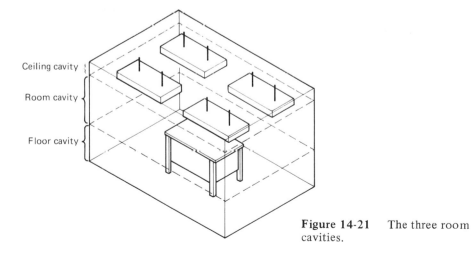

Ceiling cavity

Room cavity

Floor cavity

Figure 14-21   The three room cavities.

ceiling. Find the ceiling cavity ratio. By substituting known values in the previous formula, we have

$$\text{Ceiling cavity ratio} = \frac{5 \times 1 \, (12 + 8)}{12 \times 8}$$

$$= 1.04 \quad \text{or} \quad 1$$

For rooms composed of more than one rectangle, such as an L-shaped room, the cavity ratio is obtained by the following formula:

$$\text{Cavity ratio} = \frac{2.5 \times \text{wall area}}{\text{floor area}}$$

In calculating the ceiling cavity ratio, wall area is determined by multiplying the total linear feet of the walls by the distance between the lighting fixture plane and the ceiling cavity.

For example, an L-shaped room has the physical dimensions illustrated in Fig. 14-22. Notice that the ceiling cavity is 2 feet deep. Find the ceiling cavity ratio of this room. Find the total linear feet of the walls:

$$15 + 15 + 10 + 5 + 5 = 60 \text{ linear feet}$$

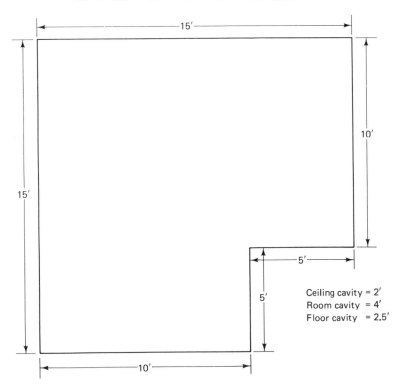

Ceiling cavity = 2'
Room cavity = 4'
Floor cavity = 2.5'

Figure 14-22   Lighting fixture suspended below ceiling in L-shaped room.

Multiply the total linear feet by the ceiling cavity depth:

$$60 \times 2 = 120 \text{ square feet}$$

Find the total floor area by dividing the room into two separate rectangles as shown in Fig. 14-23.

$$A = 5 \times 15$$
$$= 75 \text{ square feet}$$
$$B = 5 \times 10$$
$$= 50 \text{ square feet}$$
$$A + B \text{ (total floor area)} = 75 + 50$$
$$= 125 \text{ square feet}$$

Substitute these values in the formula.

$$\text{Cavity ratio} = \frac{2.5 \times \text{wall area}}{\text{floor area}}$$
$$= \frac{2.5 \times 120}{125}$$
$$= 2.4$$

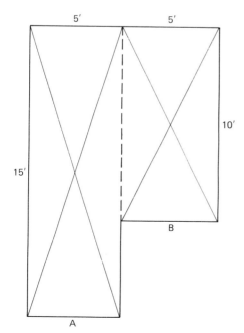

Figure 14-23 Dividing the L-shaped room into two separate rectangles.

Similarly, the floor cavity height in Fig. 14-22 is 2.5 feet. Since we already know the perimeter of the room to be 60 feet, we can multiply 2.5 by 60 and obtain a floor-cavity wall area of 150 square feet. The floor area remains the same for all three cavity calculations. Thus,

$$\text{Room cavity ratio} = \frac{2.5 \times 240}{125}$$

$$= 4.8$$

For other than rectangular rooms, the area can be calculated as required. For example, in a circular room, the cavity wall area equals height $\times 2r$ and the floor area equals r^2. Thus,

$$\text{Cavity ratio} = \frac{2.5 \times \text{height} \times 2\pi r}{\pi r^2} \quad \text{or} \quad \frac{5 \times \text{height}}{r}$$

Table 14-2 may be used to determine the cavity ratios for rectangular rooms. Using this table is usually faster than working the formulas and is usually preferred by lighting designers. Find the width of the room under consideration in the column on the far left. Then go to the next column and find the length of the room. Continue across this line until a number lines up with the column of the cavity depth (found at top of table). This number is the cavity ratio.

Effective Reflectance

Before the coefficient of utilization can be selected, the combination of ceiling and wall reflectance, as well as floor and wall reflectance, must be converted to effective ceiling or floor reflectance. The effective reflectance of the ceiling and floor cavities takes into account the effect of interflection of light among the various room surfaces. Again, charts or tables are provided for this conversion (Table 14-3). In order to find the effective reflectance, locate the column containing the known percentage of ceiling or floor reflectance and wall reflectance. Then continue down this column and read the effective reflectance opposite the appropriate cavity ratio.

As an example, assume the room illustrated in Fig. 14-21 has a ceiling reflectance of 80 percent and a wall reflectance of 50 percent. We have previously calculated the ceiling cavity ratio (1). By following the instructions given in the preceding paragraphs, we find that the effective ceiling reflectance is 66 percent.

Coefficient of Utilization

Manufacturers usually supply photometric data for their own lighting fixtures. These data contain coefficients of utilization for various surface reflections and room-cavity ratios. An accurate calculation will result in actual

TABLE 14-2

Room Dimensions		Cavity Depth																			
Width	Length	1.0	1.5	2.0	2.5	3.0	3.5	4.0	5.0	6.0	7.0	8	9	10	11	12	14	16	20	25	30
8	8	1.2	1.9	2.5	3.1	3.7	4.4	5.0	6.2	7.5	8.8	10.0	11.2	12.5	–	–	–	–	–	–	–
	10	1.1	1.7	2.2	2.8	3.4	3.9	4.5	5.6	6.7	7.9	9.0	10.1	11.3	12.4	–	–	–	–	–	–
	14	1.0	1.5	2.0	2.5	3.0	3.4	3.9	4.9	5.9	6.9	7.8	8.8	9.7	10.7	11.7	–	–	–	–	–
	20	0.9	1.3	1.7	2.2	2.6	3.1	3.5	4.4	5.2	6.1	7.0	7.9	8.8	9.6	10.5	12.2	–	–	–	–
	30	0.8	1.2	1.6	2.0	2.4	2.8	3.2	4.0	4.7	5.5	6.3	7.1	7.9	8.7	9.5	11.0	–	–	–	–
	40	0.7	1.1	1.5	1.9	2.3	2.6	3.0	3.7	4.5	5.3	5.9	6.5	7.4	8.1	8.8	10.3	11.8	–	–	–
10	10	1.0	1.5	2.0	2.5	3.0	3.5	4.0	5.0	6.0	7.0	8.0	9.0	10.0	11.0	12.0	–	–	–	–	–
	14	0.9	1.3	1.7	2.1	2.6	3.0	3.4	4.3	5.1	6.0	6.9	7.8	8.6	9.5	10.4	12.0	–	–	–	–
	20	0.7	1.1	1.5	1.9	2.3	2.6	3.0	3.7	4.5	5.3	6.0	6.8	7.5	8.3	9.0	10.5	12.0	–	–	–
	30	0.7	1.0	1.3	1.7	2.0	2.3	2.7	3.3	4.0	4.7	5.3	6.0	6.6	7.3	8.0	9.4	10.6	–	–	–
	40	0.6	0.9	1.2	1.6	1.9	2.2	2.5	3.1	3.7	4.4	5.0	5.6	6.2	6.9	7.5	8.7	10.0	12.5	–	–
	60	0.6	0.9	1.2	1.5	1.7	2.0	2.3	2.9	3.5	4.1	4.7	5.3	5.9	6.5	7.1	8.2	9.4	11.7	–	–
12	12	0.8	1.2	1.7	2.1	2.5	2.9	3.3	4.2	5.0	5.8	6.7	7.5	8.4	9.2	10.0	11.7	–	–	–	–
	16	0.7	1.1	1.5	1.8	2.2	2.5	2.9	3.6	4.4	5.1	5.8	6.5	7.2	8.0	8.7	10.2	11.6	–	–	–
	24	0.6	0.9	1.2	1.6	1.9	2.2	2.5	3.1	3.7	4.4	5.0	5.6	6.2	6.9	7.5	8.7	10.0	12.5	–	–
	36	0.6	0.8	1.1	1.4	1.7	1.9	2.2	2.8	3.3	3.9	4.4	5.0	5.5	6.0	6.6	7.8	8.8	11.0	–	–
	50	0.5	0.8	1.0	1.3	1.5	1.8	2.1	2.6	3.1	3.6	4.1	4.6	5.1	5.6	6.2	7.2	8.2	10.2	–	–
	70	0.5	0.7	1.0	1.2	1.5	1.7	2.0	2.4	2.9	3.4	3.9	4.4	4.9	5.4	5.8	6.8	7.8	9.7	12.2	–
14	14	0.7	1.1	1.4	1.8	2.1	2.5	2.9	3.6	4.3	5.0	5.7	6.4	7.1	7.8	8.5	10.0	11.4	–	–	–
	20	0.6	0.9	1.2	1.5	1.8	2.1	2.4	3.0	3.6	4.2	4.9	5.5	6.1	6.7	7.3	8.6	9.8	12.3	–	–
	30	0.5	0.8	1.0	1.3	1.6	1.8	2.1	2.6	3.1	3.7	4.2	4.7	5.2	5.8	6.3	7.3	8.4	10.5	–	–
	42	0.5	0.7	1.0	1.2	1.4	1.7	1.9	2.4	2.9	3.3	3.8	4.3	4.7	5.2	5.7	6.7	7.6	9.5	11.9	–
	60	0.4	0.7	0.9	1.1	1.3	1.5	1.8	2.2	2.6	3.1	3.5	3.9	4.4	4.8	5.2	6.1	7.0	8.8	10.9	–
	90	0.4	0.6	0.8	1.0	1.2	1.4	1.6	2.0	2.5	2.9	3.3	3.7	4.1	4.5	5.0	5.8	6.6	8.3	10.3	12.4
17	17	0.6	0.9	1.2	1.5	1.8	2.1	2.3	2.9	3.5	4.1	4.7	5.3	5.9	6.5	7.0	8.2	9.4	11.7	–	–
	25	0.5	0.7	1.0	1.2	1.5	1.7	2.0	2.5	3.0	3.5	4.0	4.5	5.0	5.5	6.0	7.0	8.0	10.0	12.5	–
	35	0.4	0.7	0.9	1.1	1.3	1.5	1.7	2.2	2.6	3.1	3.5	3.9	4.4	4.8	5.2	6.1	7.0	8.7	10.9	–
	50	0.4	0.6	0.8	1.0	1.2	1.4	1.6	2.0	2.4	2.8	3.1	3.5	3.9	4.3	4.7	5.5	6.2	7.7	9.7	11.6
	80	0.4	0.5	0.7	0.9	1.1	1.2	1.4	1.8	2.1	2.5	2.9	3.3	3.6	4.0	4.3	5.1	5.8	7.2	9.0	10.9
	120	0.3	0.5	0.7	0.8	1.0	1.2	1.3	1.7	2.0	2.3	2.7	3.0	3.4	3.7	4.0	4.7	5.4	6.7	8.4	10.1
20	20	0.5	0.7	1.0	1.2	1.5	1.7	2.0	2.5	3.0	3.5	4.0	4.5	5.0	5.5	6.0	7.0	8.0	10.0	12.5	–
	30	0.4	0.6	0.8	1.0	1.2	1.5	1.7	2.1	2.5	2.9	3.3	3.7	4.1	4.5	4.9	5.8	6.6	8.2	10.3	12.4
	45	0.4	0.5	0.7	0.9	1.1	1.3	1.4	1.8	2.2	2.5	2.9	3.3	3.6	4.0	4.3	5.1	5.8	7.2	9.1	10.9
	60	0.3	0.5	0.7	0.8	1.0	1.2	1.3	1.7	2.0	2.3	2.7	3.0	3.4	3.7	4.0	4.7	5.4	6.7	8.4	10.1
	90	0.3	0.5	0.6	0.8	0.9	1.1	1.2	1.5	1.8	2.1	2.4	2.7	3.0	3.3	3.6	4.2	4.8	6.0	7.5	9.0
	150	0.3	0.4	0.6	0.7	0.8	1.0	1.1	1.4	1.7	2.0	2.3	2.6	2.9	3.2	3.4	4.0	4.6	5.7	7.2	8.6
24	24	0.4	0.6	0.8	1.0	1.2	1.5	1.7	2.1	2.5	2.9	3.3	3.7	4.1	4.5	5.0	5.8	6.7	8.2	10.3	12.4
	32	0.4	0.5	0.7	0.9	1.1	1.3	1.5	1.8	2.2	2.6	2.9	3.3	3.6	4.0	4.3	5.1	5.8	7.2	9.0	11.0
	50	0.3	0.5	0.6	0.8	0.9	1.1	1.2	1.5	1.8	2.2	2.5	2.8	3.1	3.4	3.7	4.4	5.0	6.2	7.8	9.4
	70	0.3	0.4	0.6	0.7	0.8	1.0	1.1	1.4	1.7	2.0	2.2	2.5	2.8	3.0	3.3	3.8	4.4	5.5	6.9	8.2
	100	0.3	0.4	0.5	0.6	0.8	0.9	1.0	1.3	1.6	1.8	2.1	2.4	2.6	2.9	3.1	3.7	4.2	5.2	6.5	7.9
	160	0.2	0.4	0.5	0.6	0.7	0.8	1.0	1.2	1.4	1.7	1.9	2.1	2.4	2.6	2.8	3.3	3.8	4.7	5.9	7.1
30	30	0.3	0.5	0.7	0.8	1.0	1.2	1.3	1.7	2.0	2.3	2.7	3.0	3.3	3.7	4.0	4.7	5.4	6.7	8.4	10.0
	45	0.3	0.4	0.6	0.7	0.8	1.0	1.1	1.4	1.7	1.9	2.2	2.5	2.7	3.0	3.3	3.8	4.4	5.5	6.9	8.2
	60	0.3	0.4	0.5	0.6	0.7	0.9	1.0	1.2	1.5	1.7	2.0	2.2	2.5	2.7	3.0	3.5	4.0	5.0	6.2	7.4
	90	0.2	0.3	0.4	0.6	0.7	0.8	0.9	1.1	1.3	1.6	1.8	2.0	2.2	2.5	2.7	3.1	3.6	4.5	5.6	6.7
	150	0.2	0.3	0.4	0.5	0.6	0.7	0.8	1.0	1.2	1.4	1.6	1.8	2.0	2.2	2.4	2.8	3.2	4.0	5.0	5.9
	200	0.2	0.3	0.4	0.5	0.6	0.7	0.8	1.0	1.1	1.3	1.5	1.7	1.9	2.0	2.2	2.6	3.0	3.7	4.7	5.6
36	36	0.3	0.4	0.6	0.7	0.8	1.0	1.1	1.4	1.7	1.9	2.2	2.5	2.8	3.0	3.3	3.9	4.4	5.5	6.9	8.3
	50	0.2	0.4	0.5	0.6	0.7	0.8	1.0	1.2	1.4	1.7	1.9	2.1	2.3	2.6	2.9	3.3	3.8	4.8	5.9	7.2
	75	0.2	0.3	0.4	0.5	0.6	0.7	0.8	1.0	1.2	1.4	1.6	1.8	2.0	2.3	2.5	2.9	3.3	4.1	5.1	6.1
	100	0.2	0.3	0.4	0.5	0.6	0.7	0.8	0.9	1.1	1.3	1.5	1.7	1.9	2.1	2.3	2.6	3.0	3.8	4.7	5.7
	150	0.2	0.3	0.3	0.4	0.5	0.6	0.7	0.9	1.0	1.2	1.4	1.6	1.7	1.9	2.1	2.4	2.8	3.5	4.3	5.2
	200	0.2	0.2	0.3	0.4	0.5	0.6	0.7	0.8	1.0	1.1	1.3	1.5	1.6	1.8	2.0	2.3	2.6	3.3	4.1	4.9
42	42	0.2	0.4	0.5	0.6	0.7	0.8	1.0	1.2	1.4	1.6	1.9	2.1	2.4	2.6	2.8	3.3	3.8	4.7	5.9	7.1
	60	0.2	0.3	0.4	0.5	0.6	0.7	0.8	1.0	1.2	1.4	1.6	1.8	2.0	2.2	2.4	2.8	3.2	4.0	5.0	6.0
	90	0.2	0.3	0.3	0.4	0.5	0.6	0.7	0.9	1.0	1.2	1.4	1.6	1.7	1.9	2.1	2.4	2.8	3.5	4.4	5.2
	140	0.2	0.2	0.3	0.4	0.4	0.5	0.6	0.8	0.9	1.1	1.2	1.4	1.5	1.7	1.9	2.2	2.5	3.1	3.9	4.6
	200	0.1	0.2	0.3	0.4	0.4	0.5	0.5	0.7	0.9	1.0	1.1	1.3	1.4	1.6	1.7	2.0	2.3	2.9	3.6	4.3
	300	0.1	0.2	0.3	0.3	0.4	0.5	0.5	0.7	0.8	0.9	1.1	1.3	1.4	1.5	1.7	1.9	2.2	2.8	3.5	4.2
50	50	0.2	0.3	0.4	0.5	0.6	0.7	0.8	1.0	1.2	1.4	1.6	1.8	2.0	2.2	2.4	2.8	3.2	4.0	5.0	6.0
	70	0.2	0.3	0.3	0.4	0.5	0.6	0.7	0.9	1.0	1.2	1.4	1.5	1.7	1.9	2.0	2.4	2.7	3.4	4.3	5.1
	100	0.1	0.2	0.3	0.4	0.4	0.5	0.6	0.7	0.9	1.0	1.2	1.3	1.5	1.6	1.8	2.1	2.4	3.0	3.7	4.5
	150	0.1	0.2	0.3	0.4	0.4	0.5	0.5	0.7	0.8	0.9	1.1	1.2	1.3	1.5	1.6	1.9	2.1	2.7	3.3	4.0
	300	0.1	0.2	0.2	0.3	0.3	0.4	0.5	0.6	0.7	0.8	0.9	1.0	1.1	1.3	1.4	1.6	1.9	2.3	2.9	3.5
60	60	0.1	0.2	0.3	0.4	0.5	0.6	0.7	0.8	1.0	1.2	1.3	1.5	1.7	1.8	2.0	2.3	2.7	3.3	4.2	5.0
	100	0.1	0.2	0.3	0.3	0.4	0.5	0.5	0.7	0.8	0.9	1.1	1.2	1.3	1.5	1.6	1.9	2.1	2.7	3.3	4.0
	150	0.1	0.2	0.2	0.3	0.3	0.4	0.5	0.6	0.7	0.8	0.9	1.0	1.2	1.3	1.4	1.6	1.9	2.3	2.9	3.5
	300	0.1	0.1	0.2	0.2	0.3	0.3	0.4	0.5	0.6	0.7	0.8	0.9	1.0	1.1	1.2	1.4	1.6	2.0	2.5	3.0
75	75	0.1	0.2	0.3	0.3	0.4	0.5	0.5	0.7	0.8	0.9	1.1	1.2	1.3	1.5	1.6	1.9	2.1	2.7	3.3	4.0
	120	0.1	0.2	0.2	0.3	0.3	0.4	0.4	0.5	0.6	0.8	0.9	1.0	1.1	1.2	1.3	1.5	1.7	2.2	2.7	3.3
	200	0.1	0.1	0.2	0.2	0.3	0.3	0.4	0.5	0.5	0.6	0.7	0.8	0.9	1.0	1.1	1.3	1.5	1.8	2.3	2.7
	300	0.1	0.1	0.2	0.2	0.3	0.3	0.3	0.4	0.5	0.6	0.7	0.7	0.8	0.9	1.0	1.2	1.3	1.7	2.1	2.5
100	100	0.1	0.1	0.2	0.2	0.3	0.3	0.4	0.5	0.6	0.7	0.8	0.9	1.0	1.1	1.2	1.4	1.6	2.0	2.5	3.0
	200	0.1	0.1	0.1	0.2	0.2	0.3	0.3	0.4	0.4	0.5	0.6	0.6	0.7	0.8	0.9	1.0	1.2	1.5	1.9	2.2
	300	0.1	0.1	0.1	0.2	0.2	0.2	0.3	0.3	0.4	0.5	0.5	0.6	0.7	0.7	0.8	0.9	1.1	1.3	1.7	2.0
150	150	0.1	0.1	0.1	0.2	0.2	0.2	0.3	0.3	0.4	0.5	0.5	0.6	0.7	0.7	0.8	0.9	1.1	1.3	1.7	2.0
	300	–	0.1	0.1	0.1	0.1	0.2	0.2	0.2	0.3	0.3	0.4	0.5	0.5	0.6	0.6	0.7	0.8	1.0	1.2	1.5
200	200	–	0.1	0.1	0.1	0.1	0.2	0.2	0.2	0.3	0.3	0.4	0.5	0.5	0.6	0.6	0.7	0.8	1.0	1.2	1.5
	300	–	0.1	0.1	0.1	0.1	0.1	0.2	0.2	0.2	0.3	0.3	0.4	0.4	0.5	0.5	0.6	0.7	0.8	1.0	1.2
300	300	–	–	0.1	0.1	0.1	0.1	0.1	0.2	0.2	0.2	0.3	0.3	0.3	0.4	0.4	0.5	0.5	0.6	0.7	0.8
500	500	–	–	–	–	0.1	0.1	0.1	0.1	0.1	0.1	0.2	0.2	0.2	0.2	0.2	0.3	0.3	0.4	0.5	0.6

Source: Illuminating Engineering Society.

TABLE 14-3

% Ceiling or Floor Reflectance	90				80				70			50			30				10		
% Wall Reflectance	90	70	50	30	80	70	50	30	70	50	30	70	50	30	65	50	30	10	50	30	10
0	90	90	90	90	80	80	80	80	70	70	70	50	50	50	30	30	30	30	10	10	10
0.1	90	89	88	87	79	79	78	78	69	69	68	59	49	48	30	30	29	29	10	10	10
0.2	89	88	86	85	79	78	77	76	68	67	66	49	48	47	30	29	29	28	10	10	9
0.3	89	87	85	83	78	77	75	74	68	66	64	49	47	46	30	29	28	27	10	10	9
0.4	88	86	83	81	78	76	74	72	67	65	63	48	46	45	30	29	27	26	11	10	9
0.5	88	85	81	78	77	75	73	70	66	64	61	48	46	44	29	28	27	25	11	10	9
0.6	88	84	80	76	77	75	71	68	65	62	59	47	45	43	29	28	26	25	11	10	9
0.7	88	83	78	74	76	74	70	66	65	61	58	47	44	42	29	28	26	24	11	10	8
0.8	87	82	77	73	75	73	69	65	64	60	56	47	43	41	29	27	25	23	11	10	8
0.9	87	81	76	71	75	72	68	63	63	59	55	46	43	40	29	27	25	22	11	9	8
1.0	86	80	74	69	74	71	66	61	63	58	53	46	42	39	29	27	24	22	11	9	8
1.1	86	79	73	67	74	71	65	60	62	57	52	46	41	38	29	26	24	21	11	9	8
1.2	86	78	72	65	73	70	64	58	61	56	50	45	41	37	29	26	23	20	11	9	7
1.3	85	78	70	64	73	69	63	57	61	55	49	45	40	36	29	26	23	20	12	9	7
1.4	85	77	69	62	72	68	62	55	60	54	48	45	40	35	28	26	22	19	12	9	7
1.5	85	76	68	61	72	68	61	54	59	53	47	44	39	34	28	25	22	18	12	9	7
1.6	85	75	66	59	71	67	60	53	59	52	45	44	39	33	28	25	21	18	12	9	7
1.7	84	74	65	58	71	66	59	52	58	51	44	44	38	32	28	25	21	17	12	9	7
1.8	84	73	64	56	70	65	58	50	57	50	43	43	37	32	28	25	21	17	12	9	6
1.9	84	73	63	55	70	65	57	49	57	49	42	43	37	31	28	25	20	16	12	9	6
2.0	83	72	62	53	69	64	56	48	56	48	41	43	37	30	28	24	20	16	12	9	6
2.1	83	71	61	52	69	63	55	47	56	47	40	43	36	29	28	24	20	16	13	9	6
2.2	83	70	60	51	68	63	54	45	55	46	39	42	36	29	28	24	19	15	13	9	6
2.3	83	69	59	50	68	62	53	44	54	46	38	42	35	28	28	24	19	15	13	9	6
2.4	82	68	58	48	67	61	52	43	54	45	37	42	35	27	28	24	19	14	13	9	6
2.5	82	68	57	47	67	61	51	42	53	44	36	41	34	27	27	23	18	14	13	9	6
2.6	82	67	56	46	66	60	50	41	53	43	35	41	34	26	27	23	18	13	13	9	5
2.7	82	66	55	45	66	60	49	40	52	43	34	41	33	26	27	23	18	13	13	9	5
2.8	81	66	54	44	66	59	48	39	52	42	33	41	33	25	27	23	18	13	13	9	5
2.9	81	65	53	43	65	58	48	38	51	41	33	40	33	25	27	23	17	12	13	9	5
3.0	81	64	52	42	65	58	47	38	51	40	32	40	32	24	27	22	17	12	13	8	5
3.1	80	64	51	41	64	57	46	37	50	40	31	40	32	24	27	22	17	12	13	8	5
3.2	80	63	50	40	64	57	45	36	50	39′	30	40	31	23	27	22	16	11	13	8	5
3.3	80	62	49	39	64	56	44	35	49	39	30	39	31	23	27	22	16	11	13	8	5
3.4	80	62	48	38	63	56	44	34	49	38	29	39	31	22	27	22	16	11	13	8	5
3.5	79	61	48	37	63	55	43	33	48	38	29	39	30	22	26	22	16	11	13	8	5
3.6	79	60	47	36	62	54	42	33	48	37	28	39	30	21	26	21	15	10	13	8	5
3.7	79	60	46	35	62	54	42	32	48	37	27	38	30	21	26	21	15	10	13	8	4
3.8	79	59	45	35	62	53	41	31	47	36	27	38	29	21	26	21	15	10	13	8	4
3.9	78	59	45	34	61	53	40	30	47	36	26	38	29	20	26	21	15	10	13	8	4
4.0	78	58	44	33	61	52	40	30	46	35	26	38	29	20	26	21	15	9	13	8	4
4.1	78	57	43	32	60	52	39	29	46	35	25	37	28	20	26	21	14	9	13	8	4
4.2	78	57	43	32	60	51	39	29	46	34	25	37	28	19	26	20	14	9	13	8	4
4.3	78	56	42	31	60	51	38	28	45	34	25	37	28	19	26	20	14	9	13	8	4
4.4	77	56	41	30	59	51	38	28	45	34	24	37	27	19	26	20	14	8	13	8	4
4.5	77	55	41	30	59	50	37	27	45	33	24	37	27	19	25	20	14	8	14	8	4
4.6	77	55	40	29	59	50	37	26	44	33	24	36	27	18	25	20	14	8	14	8	4
4.7	77	54	40	29	58	49	36	26	44	33	23	36	26	18	25	20	13	8	14	8	4
4.8	76	54	39	28	58	49	36	25	44	32	23	36	26	18	25	19	13	8	14	8	4
4.9	76	53	38	28	58	49	35	25	44	32	23	36	26	18	25	19	13	7	14	8	4
5.0	76	53	38	27	57	48	35	25	43	32	22	36	26	17	25	19	13	7	14	8	4

Source: Illuminating Engineering Society.

applications if the coefficients of utilization of the actual lighting fixture under consideration are used. Table 14-4 is typical of such data. This table of coefficients is based on an effective floor-cavity reflectance of 20 percent. If it is substantially different, an adjustment is made later.

Assume that we want the coefficient of utilization for: effective ceiling-cavity reflectance (% CCR) of 0.78, wall reflectance (% WR) of 0.5, and room-cavity ratio (RCR) of 4 from Table 14-4. The effective floor-cavity reflectance (% FCR) is 18 percent and is close enough to 20 percent for our purposes. The 78 percent ceiling reflectance is also close enough to the 80 percent reflectance in Table 14-4 without adjustment.

For an 80 percent effective ceiling-cavity reflectance, a 50 percent wall reflectance, and a room cavity ratio of 4, Table 14-4 gives a coefficient of utilization of 0.43.

TABLE 14-4. Coefficients of Utilization

% CCR	80%		50%	
% FCR	20%		20%	
% WR	70%	50%	50%	30%
RCR 1	.60	.58	.54	.53
2	.56	.52	.49	.47
3	.52	.47	.45	.42
4	.48	.43	.31	.38
5	.45	.39	.37	.34
6	.42	.36	.34	.31
7	.39	.33	.31	.28
8	.36	.30	.29	.25
9	.34	.27	.26	.23
10	.31	.25	.24	.21

Source: Illuminating Engineering Society.

Correction Factor and Interpolation

If the effective floor-cavity reflectance is 18, 19, 21, or 22 percent, direct use of coefficient of utilization tables can be made. However, if the effective floor-cavity reflectance is 17 percent or below or 23 percent and above, the lighting designer may find that an adjustment is necessary. Table 14-5 lists correction factors for effective floor-cavity reflectances other than 20 percent.

To demonstrate the use of Table 14-5 and to provide an example of interpolation, assume the effective floor reflectance is 17 percent.

Table 14-5 shows that, for an 80 percent ceiling, 50 percent wall, and 4-room cavity ratio, the correction factor is 1.05. This will be applied to the

TABLE 14-5

Room Cavity Ratio	Percent Effective Ceiling-Cavity Reflectance											
	80			70			50			10		
	Percent Wall Reflectance											
	50	30	10	50	30	10	50	30	10	50	30	10
1	1.08	1.08	1.07	1.07	1.06	1.06	1.05	1.04	1.04	1.01	1.01	1.01
2	1.07	1.06	1.05	1.06	1.05	1.04	1.04	1.03	1.03	1.01	1.01	1.01
3	1.05	1.04	1.03	1.05	1.04	1.03	1.03	1.03	1.02	1.01	1.01	1.01
4	1.05	1.03	1.02	1.04	1.03	1.02	1.03	1.02	1.02	1.01	1.01	1.00
5	1.04	1.03	1.02	1.03	1.02	1.02	1.02	1.02	1.01	1.01	1.01	1.00
6	1.03	1.02	1.01	1.03	1.02	1.01	1.02	1.02	1.01	1.01	1.01	1.00
7	1.03	1.02	1.01	1.03	1.02	1.01	1.02	1.01	1.01	1.01	1.01	1.00
8	1.03	1.02	1.01	1.02	1.02	1.01	1.02	1.01	1.01	1.01	1.01	1.00
9	1.02	1.01	1.01	1.02	1.01	1.01	1.02	1.01	1.01	1.01	1.01	1.00
10	1.02	1.01	1.01	1.02	1.01	1.01	1.02	1.01	1.01	1.01	1.01	1.00

If the effective floor cavity reflectance is

30%:	Multiply the luminaire CU by the factor shown in this table.
23 to 29%:	Multiply the luminaire CU by the result obtained from interpolating between 1.00 and the factor shown in this table.
18 to 22%:	Use the luminaire CU directly; do not use this table.
11 to 17%:	Divide the luminaire CU by the result obtained from interpolating between 1.00 and the factor shown in this table.
10%:	Divide the luminaire CU by the factor shown in this table.

Source: Illuminating Engineering Society.

158

tentative coefficient of utilization of 0.43 obtained previously. The instructions in Table 14-5 require interpolation between 1 and 1.05 as follows:

1. Seventeen percent is three-tenths of the way from 20 to 10 percent.
2. A factor of 1 applies to 20 percent and 1.05 applies to 10 percent.
3. Therefore, the correction factor we want is three-tenths of the way from 1 to 1.05.
4. Multiplying 3/10 by (1.05 − 1) gives 0.3 × 0.05 = 0.015.
5. Adding 0.015 and 1 gives 1.015 as the required correction factor.

The instructions in Table 14-5 state that the lighting fixture coefficient of utilization is to be divided by this factor. Thus, 0.43 ÷ 1.015 = 0.424, and the corrected coefficient of utilization for a 17 percent effective floor-cavity reflectance will be 0.42 instead of 0.43.

Maintenance Factor

The maintenance factor takes into account the reduction in light output because of lamp aging and dirt accumulation. The appropriate maintenance factor for any given condition and lighting fixture type may be determined as follows.

Types of lighting fixtures are divided into five categories. The category for each lighting fixture in Appendix A is indicated in the fixture column. After determining the category, the maintenance factor can be read from one of the five curves for each category in Fig. 14-24. The point on the curve should be selected on the basis of the estimated number of months between cleaning of the lighting fixtures. The particular curve selected should be based on the dirt content of the atmosphere under consideration.

Number of Lamps and Lighting Fixtures Required

The number of lighting fixtures and lamps can be calculated from the following formula:

$$NF = \frac{FA \times DF}{LPF \times LPL \times CU \times MF}$$

where

NF = number of fixtures
FA = floor area
DF = desired footcandles
LPF = lamps per fixture
LPL = lumens per lamp
CU = coefficient of utilization
MF = maintenance factor

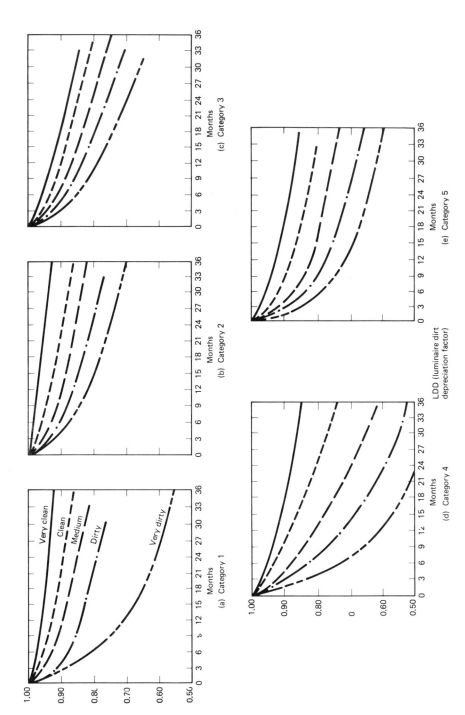

Figure 14-24 Maintenance factor curves of the five lighting fixture categories.

LDD (luminaire dirt depreciation factor)

(a) Category 1

(b) Category 2

(c) Category 3

(d) Category 4

(e) Category 5

Very clean

Clean

Medium

Dirty

Very dirty

As an example, using the room as illustrated in Fig. 14-21, assume the seeing task to be 150 footcandles. We have previously determined the floor area to be 96 square feet and the coefficient of utilization to be 0.42. The lighting fixtures that will be used will contain four 40-watt fluorescent lamps. Assume a maintenance factor of 0.7. Referring to lamp data tables, we find that this lamp has approximately 3250 initial lumens. By inserting these values in the formula, we have

$$NF = \frac{96 \times 150}{4 \times 3250 \times 0.42 \times 0.7}$$

$$= 3.8, \quad \text{or } 4 \text{ fixtures}$$

Four lighting fixtures will then be needed to obtain approximately 150 footcandles of illumination. Since the calculation did not come out even (3.8), four fixtures will give slightly more than 150 footcandles, but this slight difference is insignificant for all practical purposes.

Layout of Lighting Fixtures

Lighting fixture locations depend on the general architecture, size of bays, type of lighting fixture under consideration, and so on. To provide even distribution of illumination for an area, the permissible maximum spacing recommendations should not be exceeded. These recommendation ratios are supplied in terms of maximum spacing to mounting height. The "Spacing Not to Exceed" column of Appendix A gives maximum permissible ratios of spacing to mounting height above the work plane for the types of lighting fixture included. In most cases it is necessary to locate fixtures closer together than these maximums in order to obtain required illumination levels.

In our example, the interior dimensions of the room in Fig. 14-21 are 12 feet by 8 feet and the floor plan is shown in Fig. 14-25.

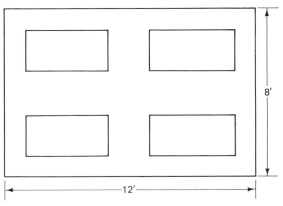

Figure 14-25   Floor plan of room in Fig. 14-21.

Adjusted Footcandles

As mentioned previously, the exact number of lighting fixtures required for this room is 3.8. However, it would be very difficult indeed to install 0.8 of a lighting fixture, so 4 fixtures were used. The amount of actual illumination then can be recomputed by the formula

$$FC = \frac{TL \times LPL \times CU \times MF}{A}$$

where

FC = footcandles
TL = total lamps
LPL = lumens per lamp
CU = coefficient of utilization
MF = maintenance factor
A = area

Thus,

$$FC = \frac{16 \times 3250 \times 0.42 \times 0.7}{96}$$

$$= 159 \text{ footcandles}$$

15

Electrical Machines

The principal means of changing electrical energy into mechanical energy or power is the electric motor, ranging in size from small fractional-horsepower, low-voltage motors to the very large high-voltage synchronous motors. Electric motors are classified according to size (horsepower); type of application; electrical characteristics; speed, starting, speed control and torque characteristics; and mechanical protection and method of cooling.

In basic terms, electric motors convert electric energy into the productive power of rotary mechanical force. This capability finds application in unlimited ways from explosionproof, water-cooled motors for underground mining to induced-draft fan motors for power generation; from adjustable-frequency drives for waste and water treatment pumping to eddy-current clutches for automobile production; from direct-current drive systems for paper production to photographic film manufacturing; from rolled-shell shaftless motors for machine tools to large outdoor motors for crude oil pipelines; from mechanical variable-speed drives for woodworking machines to complex adjustable-speed drive systems for textiles. All these and more represent the scope of electric motor participation in powering and controlling the machines and processes of industries through the world.

Before covering electric motor operating principles and their applications, certain motor terms must be understood. Some of the more basic ones are as follows:

Style number: Identifies that particular motor from any other. Manufacturers provide style numbers on the motor nameplate and in the written specifications.

Serial data code: The first letter is a manufacturing code used at the factory. The second letter identifies the month, and the last two numbers identify the year of manufacture (D78 is April 78).

Frame (fr): Specifies the shaft height and motor mounting dimensions and provides recommendations for standard shaft diameters and usable shaft extension lengths.

Service factor: A service factor (SF) is a multiplier that, when applied to the rated horsepower, indicates a permissible horsepower loading that may be carried continuously when the voltage and frequency are maintained at the value specified on the nameplate, although the motor will operate at an increased temperature rise.

NEMA service factors: Open motors only:

Hp	SF	Hp	SF	Hp	SF
1/12	1.40	1/3	1.35	1	1.15
1/8	1.40	1/2	1.25	1-1/2	1.15
1/6	1.35	3/4	1.25	2	1.15
1/4	1.35			3	1.15

1 hp, open motor at 3450 rpm, has 1.25 service factor.

Phase: Indicates whether the motor has been designed for single or three phase. It is determined by the electrical power source.

Degree C ambient: The air temperature immediately surrounding the motor. Forty degrees centigrade is the NEMA maximum ambient temperature.

Insulation class: The insulation system is chosen to ensure the motor will perform at the rated horsepower and service factor load.

Horsepower: Defines the rated output capacity of the motor. It is based on breakdown torque, which is the maximum torque a motor will develop without an abrupt drop in speed.

RPM: Revolutions per minute (speed). The RPM reading on motors is the approximate full-load speed. The speed of the motor is determined by the number of poles in the winding. A two-pole motor runs at an approximate speed of 3450 rpm. A four-pole motor runs at an approximate speed of 1725 rpm. A six-pole motor runs at an approximate speed of 1140 rpm.

Amps: Gives the amperes of current the motor draws at full load. When two values are shown on the nameplate, the motor usually has a dual voltage rating. Volts and amperes are inversely proportional; the higher the voltage the lower the amperes, and vice versa. The higher ampere value corresponds to the lower voltage rating on the nameplate. Two-speed motors will also show two ampere readings.

Hertz (cycles): Just about everything in this country is serviced by 60-hertz alternating current. Therefore, most applications will be for 60-hertz operations.

Volts: Volts is the electrical potential "pressure" for which the motor is designed. Sometimes two voltages are listed on the nameplate, like 115/230. In this case the motor is intended for use on either a 115 or 230 circuit. Special instructions are furnished for connecting the motor for each of the different voltages.

KVA code: This code letter is defined by NEMA standards to designate the locked rotor kilovolt-amperes (kVA) per horsepower of a motor. It relates to starting current and selection of fuse or circuit breaker size.

Housing: Designates the type of motor enclosure. The most common types are open and enclosed:

> *Open drip-proof* has ventilating openings so constructed that successful operation is not interfered with when drops of liquid or solid particles strike or enter the enclosure at any angle from 0 to 15° downward from the vertical.

> *Open guarded* has all openings giving direct access to live metal hazardous rotating parts so sized or shielded as to prevent accidental contact as defined by probes illustrated in the NEMA standard.

> *Totally enclosed* motors are so constructed to prevent the free exchange of air between the inside and outside of the motor casing.

> *Totally enclosed fan-cooled* motors are equipped for external cooling by means of a fan that is integral with the motor.

> *Air-over* motors must be mounted in the airstream to obtain their nameplate rating without overheating. An air-over motor may be either open or enclosed.

Explosionproof motors: These are totally enclosed designs built to withstand an explosion of gas or vapor within it, and to prevent ignition of the gas or vapor surrounding the motor by sparks or explosions that may occur within the motor casing.

Hours: Designates the duty cycle of a motor. Most fractional-horse-power motors are marked continuous for around-the-clock operation at nameplate rating in the rated ambient. Motors marked "one-half" are for half-hour ratings, and those marked "one" are for one-hour ratings.

The following terms are not found on the nameplate but are important considerations for proper motor selection.

BEARINGS

Sleeve bearings: Sleeve bearings are generally recommended for axial thrust loads of 210 pounds or less and are designed to operate in any mounting position as long as the belt pull is not against the bearing window. On light-duty applications, sleeve bearings can be expected to perform a minimum of 25,000 hours without relubrication.

Ball bearings: These are recommended where axial thrust exceeds 20 pounds. They too can be mounted in any position. Standard and general-purpose ball-bearing motors are factory lubricated and under normal conditions will require no additional lubrication for many years.

MOUNTINGS

Rigid mounting: A rectangular steel mounting plate that is welded to the motor frame or cast integral with the frame; it is the most common type of mounting.

Resilient mounting: A mounting base that is isolated from motor vibration by means of rubber rings secured to the end bells.

Flange mounting: A special end bell with a machined flange that has two or more holes through which bolts are secured. Flange mountings are commonly used on such applications as jetty pumps and oil burners.

Rotation: For single-phase motors, the standard rotation, unless otherwise noted, is counterclockwise facing the lead or opposite shaft end. All motors can be reconnected at the terminal board for opposite rotation, unless otherwise indicated.

SINGLE-PHASE MOTORS

Split-Phase Motors

Split-phase motors are fractional-horsepower units that use an auxiliary winding on the stator to aid in starting the motor until it reaches its proper rotation speed (see Fig. 15-1). This type of motor finds use in small pumps, oil burners, automatic washers, and other household appliances.

In general, the split-phase motor consists of a housing, a laminated iron-core stator with embedded windings forming the inside of the cylindrical housing, a rotor made up of copper bars set in slots in an iron core and connected to each other by copper rings around both ends of the core, plates that are bolted to the housing and contain the bearings that support the rotor shaft, and a centrifugal switch inside the housing. This type of rotor is often called a *squirrel cage* rotor since the configuration of the copper bars resembles an actual cage. These motors have no windings as such, and a centrifugal switch is provided to open the circuit to the starting winding when the motor reaches running speed.

To understand the operation of a split-phase motor, look at the wiring diagram in Fig. 15-1. Current is applied to the stator windings, both the main winding and the starting winding, which is in parallel with it through the centrifugal switch. The two windings set up a rotating magnetic field, and this field sets up a voltage in the copper bars of the squirrel-cage rotor. Because these bars are shortened at the ends of the rotor, current flows through the rotor bars. The current-carrying rotor bars then react with the magnetic field to produce motor action. When the rotor is turning at the proper speed, the centrifugal switch cuts out the starting winding since it is no longer needed.

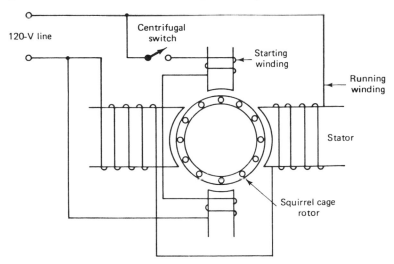

Figure 15-1   Schematic diagram of a standard split-phase ac motor.

Capacitor Motors

Capacitor motors are single-phase ac motors ranging in size from fractional horsepower (hp) to perhaps as high as 15 hp. This type of motor is widely used in all types of single-phase applications such as powering machine shop tools (lathes, drill presses, etc.), air compressors, refrigerators, and the like. This type of motor is similar in construction to the split-phase motor, except a capacitor is wired in series with the starting winding, as shown in Fig. 15-2.

The capacitor provides higher starting torque, with lower starting current, than does the split-phase motor, and although the capacitor is sometimes mounted inside the motor housing, it is more often mounted on top of the motor, encased in a metal compartment.

In general, two types of capacitor motors are in use: the capacitor-start motor and the capacitor start-and-run motor. As the name implies, the former utilizes the capacitor only for starting; it is disconnected from the circuit once the motor reaches running speed, or at about 75 percent of the motor's full speed. Then the centrifugal switch opens to cut the capacitor out of the circuit.

The capacitor start-and-run motor keeps the capacitor and starting winding in parallel with the running winding, providing a quiet and smooth operation at all times.

Capacitor split-phase motors require the least maintenance of all single-phase motors, but they have a very low starting torque, making them unsuitable for many applications. Its high maximum torque, however, makes it

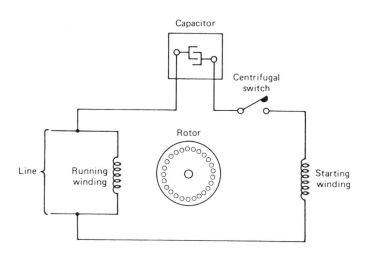

Figure 15-2   Wiring diagram of a typical capacitor-start motor.

especially useful for such tools as floor sanders or in grinders where momentary overloads due to excessive cutting pressure are experienced. It is also used quite frequently for slow-speed direct-connected fans.

Repulsion-Type Motors

Repulsion-type motors are divided into several groups, including (1) repulsion-start, induction-run motors, (2) repulsion motors, and (3) repulsion-induction motors. The repulsion-start, induction-run motor is of the single-phase type, ranging in size from about 1/10 hp to as high as 20 hp. It has high starting torque and a constant-speed characteristic, which makes it suitable for such applications as commercial refrigerators, compressors, pumps, and similar applications requiring high starting torque.

The repulsion motor is distinguished from the repulsion-start, induction-run motor by the fact that it is made exclusively as a brush-riding type and does not have any centrifugal mechanism. Therefore, this motor both starts and runs on the repulsion principle. This type of motor has high starting torque and a variable-speed characteristic. It is reversed by shifting the brush holder to either side of the neutral position. Its speed can be decreased by moving the brush holder farther away from the neutral position.

The repulsion-induction motor combines the high starting torque of the repulsion-type and the good speed regulation of the induction motor. The stator of this motor is provided with a regular single-phase winding, while the rotor winding is similar to that used on a dc motor. When starting, the changing single-phase stator flux cuts across the rotor windings and induces currents in them; thus, when flowing through the commutator, a continuous repulsive action on the stator poles is present.

This motor starts as a straight repulsion-type and accelerates to about 75 percent of normal full speed when a centrifugally operated device connects all the commutator bars together and converts the winding to an equivalent squirrel-cage type. The same mechanism usually raises the brushes to reduce noise and wear. Note that, when the machine is operating as a repulsion-type, the rotor and stator poles reverse at the same instant, and that the current in the commutator and brushes is ac.

This type of motor will develop four to five times normal full-load torque and will draw about three times normal full-load current when starting with full-line voltage applied. The speed variation from no load to full load will not exceed 5 percent of normal full-load speed.

The repulsion-induction motor is used to power air compressors, refrigeration, pumps, meat grinders, small lathes, small conveyors, stokers, and the like. In general, this type of motor is suitable for any load that requires a high starting torque and constant-speed operation. Most motors of this type are less than 5 hp.

Universal Motors

This type of motor is a special adaptation of the series-connected dc motor, and it gets its name "universal" from the fact that it can be connected on either ac or dc and operate the same. All are single-phase motors for use on 120 or 240 volts.

In general, the universal motor contains field windings on the stator within the frame, an armature with the ends of its windings brought out to a commutator at one end, and carbon brushes that are held in place by the motor's end plate, allowing them to have a proper contact with the commutator.

When current is applied to a universal motor, either ac or dc, the current flows through the field coils and the armature windings in series. The magnetic field set up by the field coils in the stator react with the current-carrying wires on the armature to produce rotation.

Universal motors are used on such household appliances as sewing machines, vacuum cleaners, and electric fans.

Shaded-Pole Motor

A shaded-pole motor is a single-phase induction motor provided with an uninsulated and permanently short-circuited auxiliary winding displaced in magnetic position from the main winding. The auxiliary winding is known as the shading coil and usually surrounds from one-third to one-half of the pole (see Fig. 15-3). The main winding surrounds the entire pole and may consist of one or more coils per pole.

Applications for this motor include small fans, timing devices, relays, radio dials, or any constant-speed load not requiring high starting torque.

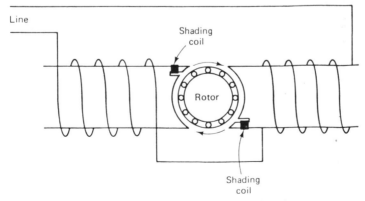

Figure 15-3   Wiring diagram of a shaded-pole motor.

POLYPHASE MOTORS

Three-phase motors offer extremely efficient and economical application and are usually the preferred type for commercial and industrial applications when three-phase service is available. In fact, the great bulk of motors sold are standard ac three-phase motors. These motors are available in ratings from fractional horsepower up to thousands of horsepower in practically every standard voltage and frequency. In fact, there are few applications for which the three-phase motor cannot be put to use.

Three-phase motors are noted for their relatively constant speed characteristic and are available in designs giving a variety of torque characteristics; that is, some have a high starting torque and others a low starting torque. Some are designed to draw a normal starting current, others a high starting current.

A typical three-phase motor is shown in Fig. 15-4. Note that the three main parts are the stator, rotor, and end plates. It is very similar in construction to conventional split-phase motors except that the three-phase motor has no centrifugal switch.

The stator shown in Fig. 15-5 consists of a steel frame and a laminated iron core and winding formed of individual coils placed in slots. The rotor may be a squirrel-cage or wound-rotor type. Both types contain a laminated core pressed onto a shaft. The squirrel-cage rotor is shown in Fig. 15-6 and is similar to a split-phase motor. The wound rotor is shown in Fig. 15-7 and has a winding on the core that is connected to three slip rings mounted on the shaft.

The end plates or brackets are bolted to each side of the stator frame and contain the bearings in which the shaft revolves. Either ball bearings or sleeve bearings are used.

Induction motors, both single-phase and polyphase, get their name from the fact that they utilize the principle of electromagnetic induction. An induction motor has a stationary part, or stator, with windings connected to the ac supply, and a rotation part, or rotor, which contains coils or bars.

Figure 15-4   Totally enclosed, fan-cooled motor.

Figure 15-5   Stator consisting of a steel frame, laminated iron core, and a winding formed of individual coils placed in slots.

Figure 15-6   A squirrel-cage rotor. (Courtesy Westinghouse.)

Figure 15-7   Typical wound rotor.

There is no electrical connection between the stator and rotor. The magnetic field produced in the stator windings induces a voltage in the rotor coils or bars.

Since the stator windings act in the same way as the primary winding of a transformer, the stator of an induction motor is sometimes called the *primary*. Similarly, the rotor is called the *secondary* because it carries the induced voltage in the same way as the secondary of a transformer.

The magnetic field necessary for induction to take place is produced by the stator windings. Therefore, the induction-motor stator is often called the *field* and its windings are called *field windings*.

The terms primary and secondary relate to the electrical characteristics and the terms stator and rotor to the mechanical features of induction motors.

The rotor transfers the rotating motion to its shaft, and the revolving shaft drives a mechanical load or a machine, such as a pump, spindle, or clock.

Commutator segments, which are essential parts of dc motors, are not needed on induction motors. This simplifies greatly the design and the maintenance of induction motors as compared to dc motors.

The turning of the rotor in an induction motor is due to induction. The rotor, or secondary, is not connected to any source of voltage. If the magnetic field of the stator, or primary, revolves, it will induce a voltage in the rotor, or secondary. The magnetic field produced by the induced voltage acts in such a way that it makes the secondary follow the movement of the primary field.

The stator, or primary, of the induction motor does not move physically. The movement of the primary magnetic field must thus be achieved electrically. A rotating magnetic field is made possible by a combination of two or more ac voltages that are out of phase with each other and applied to the stator coils. Direct current will not produce a rotating magnetic field. In three-phase induction motors, the rotating magnetic field is obtained by applying a three-phase system to the stator windings.

The direction of rotation of the rotor in an ac motor is the same as that of its rotating magnetic field. In a three-phase motor the direction can be reversed by interchanging the connections of any two supply leads. This interchange will reverse the sequence of phases in the stator, the direction of the field rotation, and therefore the direction of rotor rotation.

SYNCHRONOUS MOTORS

A synchronous polyphase motor has a stator constructed in the same way as the stator of a conventional induction motor. The iron core has slots into which coils are wound, which are also arranged and connected in the same way as the stator coils of the induction motor. These are in turn grouped to form a three-phase connection, and the three free leads are connected to a three-phase source. Frames are equipped with air ducts, which aid the cooling of the windings, and coil guards protect the winding from damage.

The rotor of a synchronous motor carries poles that project toward the armature; they are called *salient poles*. The coils are wound on laminated pole bodies and connected to slip rings on the shaft. A squirrel-cage winding for starting the motor is embedded in the pole faces.

The pole coils are energized by direct current, which is usually supplied by a small dc generator called the *exciter*. This exciter may be mounted directly on the shaft to generate dc voltage, which is applied through brushes to slip rings. On low-speed synchronous motors, the exciter is normally belted or of a separate high-speed motor-driven type.

The dimensions and construction of synchronous motors vary greatly, depending on the rating of the motors. However, synchronous motors for industrial power applications are rarely built for less than 25 hp or so. In fact, most are 100 hp or more. All are polyphase motors when built in this size. Vertical and horizontal shafts with various bearing arrangements and various enclosures cause wide variations in the appearance of the synchronous motor.

Synchronous motors are used in electrical systems where there is need for improvement in power factor or where low power factor is not desirable. This type of motor is especially adapted to heavy loads that operate for long periods of time without stopping, such as for air compressors, pumps, ship propulsion, and the like.

The construction of the synchronous motor is well adapted for high voltages, as it permits good insulation. Synchronous motors are frequently used at 2300 volts or more. Its efficient slow-running speed is another advantage.

DIRECT-CURRENT MOTORS

A direct-current motor is a machine for converting dc electrical energy into rotating mechanical energy. The principle underlying the operation of a dc motor is called *motor action* and is based on the fact that, when a wire carrying current is placed in a magnetic field, a force is exerted on the wire, moving it through the magnetic field. There are three elements to motor action as it takes place in a dc motor:

1. Many coils of wire are wound on a cylindrical rotor or armature on the shaft of the motor.
2. A magnetic field necessary for motor action is created by placing fixed electromagnetic poles around the inside of the cylindrical motor housing. When current is passed through the fixed coils, a magnetic field is set up without the housing. Then, when the armature is placed inside the motor housing, the wires of the armature coils will be situated in the field of magnetic lines of force set up by the electromagnetic poles arranged around the stator. The stationary cylindrical part of the motor is called the stator.
3. The shaft of the armature is free to rotate because it is supported at both ends of bearing brackets. Freedom of rotation is assured by providing clearance between the rotor and the faces of the magnetic poles.

Shunt-wound DC Motors

In this type of motor, the strength of the field is not affected appreciably by a change in the load, so a relatively constant speed is obtainable. This type of motor may be used for the operation of machines that require an approximate constant speed and impose low starting torque and light overload on the motor.

Series-wound DC Motors

In motors of this type, any increase in load results in more current passing through the armature and the field windings. As the field is strengthened by this increased current, the motor speed decreases. Conversely, as the load is decreased, the field is weakened and the speed increases, and at very light loads speed may become excessive. For this reason, series-wound motors are usually directly connected or geared to the load to prevent *runaway*. The increase in armature current with an increasing load produces increased torque, so the series-wound motor is particularly suited to heavy starting duty and where severe overloads may be expected. Its speed may be adjusted by means of a variable resistance placed in series with the motor, but due to variation with load, the speed cannot be held at any constant value. This variation of speed with load becomes greater as the speed is reduced. Use of this motor is normally limited to traction and lifting service.

Compound-wound Motors

In this type of motor, the speed variation due to the load changes is much less than in the series-wound motor, but greater than in the shunt-wound motor. It also has a greater starting torque than the shunt-wound motor and is able to withstand heavier overloads. However, it has a narrower adjustable-speed range. Standard motors of this type have a cumulative-compound winding, the differential-compound winding being limited to special applications. They are used where the starting load is very heavy or where the load changes suddenly and violently, as with reciprocating pumps, printing presses, and punch presses.

Brushless DC Motors

The brushless dc motor was developed to eliminate commutator problems in missiles and spacecraft in operation above the earth's atmosphere. Two general types of brushless motors are in use: the inverter-induction motor and a dc motor with an electronic commutator.

The inverter-induction motor uses an inverter that uses the motor windings as the usual filter. The operation is square wave, and the combined efficiencies of the inverter and induction motor are at least as high as for a dc motor alone. In all cases, the motors must be designed for low starting current or else the inverter must be designed to saturate so that starting current is limited; otherwise, the transistors or silicon-controlled rectifiers in the inverter will be overloaded.

MOTOR ENCLOSURES

Electric motors differ in construction and appearance, depending on the type of service for which they are to be used. Open and closed frames are quite common. In the former enclosure, the motor's parts are covered for protection, but the air can freely enter the enclosure. Further designations for this type of enclosure include dripproof, weather-protected, and splash-proof.

Totally enclosed motors, such as the one shown in Fig. 15-4, have an airtight enclosure. They may be fan cooled or self-ventilated. An enclosed motor equipped with a fan has the fan as an integral part of the machine, but external to the enclosed parts. In the self-ventilated enclosure, no external means of cooling is provided.

The type of enclosure to use will depend on the ambient and surrounding conditions. In a drip-proof machine, for example, all ventilating openings are so constructed that drops of liquid or solid particles falling on the machine at an angle of not greater than $15°$ from the vertical cannot enter the machine, even directly or by striking and running along a horizontal or inclined surface of the machine. The application of this machine would lend itself to areas where liquids are processed.

An open motor having all air openings that give direct access to live or rotating parts, other than the shaft, limited in size by the design of the parts or by screen to prevent accidental contact with such parts is classified as a drip-proof, fully guarded machine. In such enclosures, openings shall not permit the passage of a cylindrical rod 1/2 inch in diameter, except where the distance from the guard to the live rotating parts is more than 4 inches, in which case the openings shall not permit the passage of a cylindrical rod 3/4 inch in diameter.

There are other types of drip-proof machines for special applications such as externally ventilated and pipe ventilated, which as the names imply are either ventilated by a separate motor-driven blower or cooled by ventilating air from inlet ducts or pipes.

An enclosed motor whose enclosure is designed and constructed to withstand an explosion of a specified gas or vapor that may occur within the motor and to prevent the ignition of this gas or vapor surrounding the machine is designated "explosionproof" (XP) motors.

Hazardous atmospheres (requiring XP enclosures) of both a gaseous and dusty nature are classified by the NE Code as follows:

- Class I, Group A: atmospheres containing acetylene.
- Class I, Group B: atmospheres containing hydrogen gases or vapors of equivalent hazards such as manufactured gas.
- Class I, Group C: atmospheres containing ethyl ether vapor.

- Class I, Group D: atmospheres containing gasoline, petroleum, naphtha, alcohols, acetone, lacquer-solvent vapors, and natural gas.
- Class II, Group E: atmospheres containing metal dust.
- Class II, Group F: atmospheres containing carbon-black, coal, or coke dust.
- Class II, Group G: atmospheres containing grain dust.

The proper motor enclosure must be selected to fit the particular atmospheres. However, explosionproof equipment is not generally available for Class I, Groups A and B, and it is therefore necessary to isolate motors from the hazardous area.

MOTOR TYPE

The type of motor will determine the electrical characteristics of the design. NEMA-designated designs for polyphase motors are given in Table 15-1.

An A motor is a three-phase, squirrel-cage motor designed to withstand full-voltage starting with locked rotor current higher than the values for a B motor and having a slip at rated load of less than 5 percent.

TABLE 15-1

NEMA Design	Starting Torque	Starting Current	Breakdown Torque	Full-Load Slip
A	Normal	Normal	High	Low
B	Normal	Low	Medium	Low
C	High	Low	Normal	Low
D	Very high	Low	—	High

A B motor is a three-phase, squirrel-cage motor designed to withstand full-voltage starting and developing locked rotor and breakdown torques adequate for general application, and having a slip at rated load of less than 5 percent.

A C motor is a three-phase, squirrel-cage motor designed to withstand full-voltage starting, developing locked rotor torque for special high-torque applications, and having a slip at rated load of less than 5 percent.

Design D is also a three-phase, squirrel-cage motor designed to withstand full-voltage starting, developing 275 percent locked rotor torque, and having a slip at rated load of 5 percent or more.

SELECTION OF ELECTRIC MOTORS

Each type of motor has its particular field of usefulness. Because of its simplicity, economy, and durability, the induction motor is more widely used for industrial purposes than any other type of ac motor, especially if a high-speed drive is desired.

If ac power is available, all drives requiring constant speed should use squirrel-cage induction or synchronous motors on account of their ruggedness and lower cost. Drives requiring varying speeds, such as fans, blowers, or pumps, may be driven by wound-rotor induction motors. However, if there are machine tools or other machines requiring adjustable speed or a wide range of speed control, it will probably be desirable to install dc motors on such machines and supply them from the ac system by motor-generator sets or electronic rectifiers.

Practically all constant-speed machines may be driven by ac squirrel-cage motors because they are made with a variety of speed and torque characteristics. When large motors are required or when power supply is limited, the wound-rotor motor is used even for driving constant-speed machines. A wound-rotor motor, with its controller and resistance, can develop full-load torque at starting with not more than full-load torque at starting, depending on the type of motor and the starter used.

For varying-speed service, wound-rotor motors with resistance control are used for fans, blowers, and other apparatus for continuous duty, and for cranes, hoists, and other installations for intermittent duty. The controller and resistors must be properly chosen for the particular application.

Synchronous motors may be used for practically any constant-speed drive requiring about 100 hp or over.

Cost is an important consideration where more than one type of ac motor is applicable. The squirrel-cage motor is the least expensive ac motor of the three types considered and requires very little control equipment. The wound-rotor motor is more expensive and requires additional secondary control. The synchronous motor is even more expensive and requires a source of dc excitation, as well as special synchronizing control to apply the dc power at the correct instant. When very large machines are involved, as, for example, 1000 hp or over, the cost picture may change considerably and should be checked on an individual basis.

The various types of single-phase ac motors and universal motors are used very little in industrial applications, since polyphase ac or dc power is generally available. When such motors are used, however, they are usually built into the equipment by the machinery builder, as in portable tools, office machinery, and other equipment. These motors are, as a rule, especially designed for the particular machine with which they are used.

16

Motor Control

Electric motors provide one of the principal sources of power for driving household appliances, modern machine tools, and other industrial equipment. Every motor in use, however, must be controlled, if only to start and stop it, before it becomes of any value.

Motor controllers cover a wide range of types and sizes, from a simple toggle switch to a complex system with such components as relays, timers, and switches. The common function, however, is the same in any case, that is, to control some operation of an electric motor. A motor controller will include some or all of the following functions:

- Starting and stopping
- Overload protection
- Overcurrent protection
- Reversing
- Changing speed
- Jogging
- Plugging
- Sequence control
- Pilot light indication

The controller can also provide the control for auxiliary equipment such as brakes, clutches, solenoids, heaters, and signals, and may be used to control a single motor or a group of motors.

The term *motor starter* is often used in the electrical industry and means practically the same thing as *controller.* Strictly, a motor starter is the simplest form of controller and is capable of starting and stopping the motor and providing it with overload protection.

MANUAL STARTER

A manual starter is a motor controller whose contact mechanism is operated by a mechanical linkage from a toggle handle or push button, which is in turn operated by hand. A thermal unit and direct-acting overload mechanism provide motor running overload protection. Basically, a manual starter is an ON-OFF switch with overload relays.

Manual starters are used mostly on small machine tools, fans and blowers, pumps, compressors, and conveyors. They have the lowest cost of all motor starters, have a simple mechanism, and provide quiet operation with no ac magnet hum. The contacts, however, remain closed and the lever stays in the ON position in the event of a power failure, causing the motor to automatically restart when the power returns. Therefore, low-voltage protection and low-voltage release are not possible with these manually operated starters. However, this action is an advantage when the starter is applied to motors that run continuously.

Fractional-horsepower manual starters are designed to control and provide overload protection for motors of 1 hp or less on 120- or 240-volt single-phase circuits. They are available in single- and two-pole versions and are operated by a toggle handle on the front. When a serious overload occurs, the thermal unit *trips* to open the starter contacts, disconnecting the motor from the line. The contacts cannot be reclosed until the overload relay has been reset by moving the handle to the full OFF position, after allowing about 2 minutes for the thermal unit to cool. The open-type starter will fit into a standard outlet box and can be used with a standard flush plate. The compact construction of this type of device makes it possible to mount it directly on the driven machinery and in various other places where the available space is small.

Manual motor starting switches provide ON-OFF control of single- or three-phase ac motors where overload protection is not required or is separately provided. Two- or three-pole switches are available with ratings up to 10 hp, 600 volts, three phase. The continuous current rating is 30 amperes at 250 volts maximum and 20 amperes at 600 volts maximum. The toggle operation of the manual switch is similar to the fractional-horsepower starter, and typical applications of the switch include small machine tools, pumps, fans, conveyors, and other electrical machinery that have separate motor protection. They are particularly suited to switch nonmotor loads, such as resistance heaters.

The integral horsepower manual starter is available in two- and three-pole versions to control single-phase motors up to 5 hp and polyphase motors up to 10 hp, respectively.

Two-pole starters have one overload relay and three-pole starters usually have three overload relays. When an overload relay trips, the starter mechanism unlatches, opening the contacts to stop the motor. The contacts cannot be reclosed until the starter mechanism has been reset by pressing the STOP button or moving the handle to the RESET position, after allowing time for the thermal unit to cool.

Integral horsepower manual starters with low-voltage protection prevent automatic start-up of motors after a power loss. This is accomplished with a continuous-duty solenoid, which is energized whenever the line-side voltage is present. If the line voltage is lost or disconnected, the solenoid de-energizes, opening the starter contacts. The contacts will not automatically close when the voltage is restored to the line. To close the contacts, the device must be manually reset. This manual starter will not function unless the line terminals are energized. This is a safety feature that can protect personnel or equipment from damage and is used on such equipment as conveyors, grinders, metal-working machinery, mixers, woodworking machinery, and wherever standards require low-voltage protection.

MAGNETIC CONTROLLERS

Magnetic motor controllers use electromagnetic energy for closing switches. The electromagnet consists of a coil of wire placed on an iron core. When current flows through the coil, the iron of the magnet becomes magnetized and attracts the iron bar, called the *armature.* An interruption of the current flow through the coil of wire causes the armature to drop out due to the presence of an air gap in the magnetic circuit.

Line-voltage magnetic motor starters are electromechanical devices that provide a safe, convenient, and economic means for starting and stopping motors, and they have the advantage of being controlled remotely. The great bulk of motor controllers are of this type. Therefore, the operating principles and applications of magnet motor controllers should be fully understood.

In the construction of a magnetic controller, the armature is mechanically connected to a set of contacts so that, when the armature moves to its closed position, the contacts also close. When the coil has been energized and the armature has moved to the closed position, the controller is said to be *picked up* and the armature is *seated* or *sealed-in*. Some of the magnet and armature assemblies in current use are as follows:

1. *Clapper type:*   In this type, the armature is hinged. As it pivots to seal in, the movable contacts close against the stationary contacts.

2. *Vertical action:*   The action is a straight line motion with the armature and contacts being guided so that they move in a vertical plane.

3. *Horizontal action:*   Both armature and contacts move in a straight line through a horizontal plane.

4. *Bell crank:*   A bell crank lever transforms the vertical action of the armature into a horizontal contact motion. The shock of armature pickup is not transmitted to the contacts, resulting in minimum contact bounce and longer contact life.

The magnetic circuit of a controller consists of the magnet assembly, the coil, and the armature. It is so named from a comparison with an electrical circuit. The coil and the current flowing in it cause magnetic flux to be set up through the iron in a similar manner to a voltage causing current to flow through a system of conductors. The changing magnetic flux produced by alternating currents results in a temperature rise in the magnetic circuit. The heating effect is reduced by laminating the magnet assembly and armature. By placing a coil of many turns of wire around a soft iron core, the magnetic flux set up by the energized coil tends to be concentrated; therefore, the magnetic field effect is strengthened. Since the iron core is the path of least resistance to the flow of the magnetic lines of force, magnetic attraction will concentrate according to the shape of the magnet.

The magnetic assembly is the stationary part of the magnetic circuit. The coil is supported by and surrounds part of the magnet assembly in order to induce magnetic flux into the magnetic circuit.

The armature is the moving part of the magnetic circuit. When it has been attracted into its sealed-in position, it completes the magnetic circuit. To provide maximum pull and to help ensure quietness, the faces of the armature and the magnetic assembly are ground to a very close tolerance.

When a controller's armature has sealed-in, it is held closely against the magnet assembly. However, a small gap is always deliberately left in the iron circuit. When the coil becomes de-energized, some magnetic flux (residual magnetism) always remains, and if it were not for the gap in the iron circuit, the residual magnetism might be sufficient to hold the armature in the sealed-in position.

The shaded-pole principle is used to provide a time delay in the decay of flux in dc coils, but it is used more frequently to prevent a chatter and wear in the moving parts of ac magnets. A shading coil is a single turn of conducting material mounted in the face of the magnet assembly or armature. The alternating main magnetic flux induces currents in the shading coil, and these currents set up auxiliary magnetic flux that is out of phase from the main flux. The auxiliary flux produces a magnetic pull out of phase from the pull due to the main flux, and this keeps the armature sealed-in when the main flux falls to zero (which occurs 120 times per second with 60-cycle ac). Without the shading coil, the armature would tend to open each time the main flux goes through zero. Excessive noise, wear on magnet faces, and heat would result.

Figure 16-1 shows an exaggerated view of a pole face with a copper band or short-circuited coil of low resistance connected around a portion of the pole tip. When the flux is increasing in the pole from left to right, the induced current in the coil is in a clockwise direction.

The magnetomotive force produced by the coil opposes the direction of the flux of the main field. Therefore, the flux density in the shaded portion of the iron will be considerably less, and the flux density in the unshaded portion of the iron will be more than would be the case without the shading coil.

Figure 16-2 shows the pole with the flux still moving from left to right but decreasing in value. Now the current in the coil is in a counterclockwise direction. The magnetomotive force produced by the coil is in the same direction as the main unshaded portion but less than it would be without the shading coil. Consequently, if the electric circuit of a coil is opened, the current decreases rapidly to zero, but the flux decreases much more slowly due to the action of the shading coil.

Electrical ratings for ac magnetic contactors and starters are shown in Fig. 16-3.

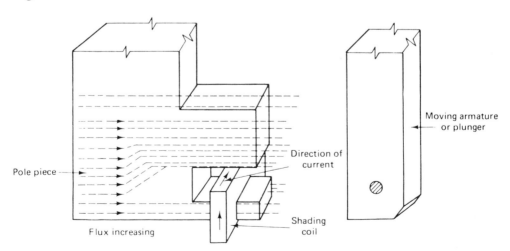

Figure 16-1   Section of pole face with current in clockwise direction.

OVERLOAD PROTECTION

Overload protection for an electric motor is necessary to prevent burnout and to ensure maximum operating life. Electric motors will, if permitted, operate at an output of more than rated capacity. Conditions of motor overload may be caused by an overload on driven machinery, by a low line voltage, or by an open line in a polyphase system, which results in single-phase operation. Under any condition of overload, a motor draws excessive current that causes overheating. Since motor winding insulation deteriorates when subjected to overheating, there are established limits on motor operating temperatures. To protect a motor from overheating, overload relays are employed on a motor control to limit the amount of current drawn. This is overload protection, or running protection.

The ideal overload protection for a motor is an element with current-sensing properties very similar to the heating curve of the motor (see Fig. 16-4), which would act to open the motor circuit when full-load current is exceeded. The operation of the protective device should be such that the motor is allowed to carry harmless overloads, but is quickly removed from the line when an overload has persisted too long.

Fuses are not designed to provide overload protection. Their basic function is to protect against short circuits (overcurrent protection). Motors draw a high inrush current when starting and conventional single-element fuses have no way of distinguishing between this temporary and harmless inrush current and a damaging overload. Such fuses, chosen on the basis of motor full-load current, would blow every time the motor is started. On the other

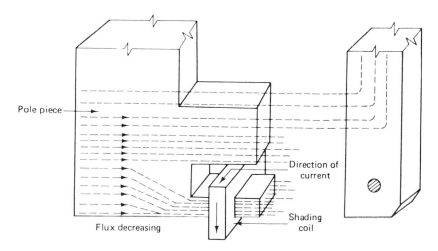

Figure 16-2   Section of pole face with current in counterclockwise direction.

NEMA Size	Load Volts	Max HP Rating — Nonplugging & Nonjogging Duty: Single Phase	Poly-Phase	Max HP Rating — Plugging & Jogging Duty †: Single Phase	Poly-Phase	Continuous Current Rating, Amperes — 600 Volt Max.	Service-Limit Current Rating, Amperes ★	Tungsten & Infrared Lamp Load, Amperes — 250 Volts Max. ★	Resistance Heating Loads, KW — other than Infrared Lamp Loads ‡: Single Phase	Poly-Phase	KVA Switching Transformer, Inrush ≤ 20× Peak: Single Phase	Poly-Phase	KVA Switching Transformer, Inrush Over 20–40× Peak: Single Phase	Poly-Phase	3 Phase Rating for Switching Capacitors ● KVAR
00	115	⅓	⋯	⋯	⋯	9	11	5	⋯	⋯	⋯	⋯	⋯	⋯	⋯
	200	⋯	1½	⋯	⋯	9	11	5	⋯	⋯	⋯	⋯	⋯	⋯	⋯
	230	1	1½	⋯	⋯	9	11	5	⋯	⋯	⋯	⋯	⋯	⋯	⋯
	380	⋯	1½	⋯	⋯	9	11	⋯	⋯	⋯	⋯	⋯	⋯	⋯	⋯
	460	⋯	2	⋯	⋯	9	11	⋯	⋯	⋯	⋯	⋯	⋯	⋯	⋯
	575	⋯	2	⋯	⋯	9	11	⋯	⋯	⋯	⋯	⋯	⋯	⋯	⋯
0	115	1	⋯	½	⋯	18	21	10	⋯	⋯	0.6	⋯	0.3	⋯	⋯
	200	⋯	3	⋯	1½	18	21	10	⋯	⋯	⋯	⋯	⋯	⋯	⋯
	230	2	3	1	1½	18	21	10	⋯	⋯	1.2	⋯	0.6	⋯	⋯
	380	⋯	5	⋯	⋯	18	21	⋯	⋯	⋯	⋯	⋯	⋯	⋯	⋯
	460	⋯	5	⋯	2	18	21	⋯	⋯	⋯	2.4	1.8	1.2	0.9	⋯
	575	⋯	5	⋯	2	18	21	⋯	⋯	⋯	3.0	2.1	1.5	1.0	⋯
1	115	2	⋯	1	⋯	27	32	15	3	5	1.2	⋯	0.6	⋯	⋯
	200	⋯	7½	⋯	3	27	32	15	⋯	9.1	⋯	⋯	⋯	⋯	⋯
	230	3	7½	2	3	27	32	15	6	10	2.4	⋯	1.2	⋯	⋯
	380	⋯	⋯	⋯	⋯	27	32	⋯	⋯	16.5	⋯	⋯	⋯	⋯	⋯
	460	⋯	10	⋯	5	27	32	⋯	12	20	4.9	3.6	2.5	1.8	⋯
	575	⋯	10	⋯	5	27	32	⋯	15	25	6.2	4.3	3.1	2.1	⋯
1P	115	3	⋯	1½	⋯	36	42	24	⋯	⋯	⋯	⋯	⋯	⋯	⋯
	230	5	⋯	3	⋯	36	42	24	⋯	⋯	⋯	⋯	⋯	⋯	⋯
2	115	3	⋯	2	⋯	45	52	30	5	8.5	2.1	⋯	1.0	⋯	⋯
	200	⋯	10	⋯	7½	45	52	30	⋯	15	⋯	⋯	⋯	⋯	⋯
	230	7½	15	5	10	45	52	30	10	17	4.1	⋯	2.1	⋯	8
	380	⋯	⋯	⋯	⋯	45	52	⋯	⋯	28	⋯	⋯	⋯	⋯	⋯
	460	⋯	25	⋯	15	45	52	⋯	20	34	8.3	6.3	4.2	3.1	16
	575	⋯	25	⋯	15	45	52	⋯	25	43	10.0	7.2	5.2	3.6	20
3	115	7½	⋯	⋯	⋯	90	104	60	10	17	4.1	⋯	2.0	⋯	⋯
	200	⋯	25	⋯	15	90	104	60	⋯	31	⋯	⋯	⋯	⋯	⋯
	230	15	30	⋯	20	90	104	60	20	34	8.1	⋯	4.1	⋯	27
	380	⋯	⋯	⋯	⋯	90	104	⋯	⋯	56	⋯	⋯	⋯	⋯	⋯
	460	⋯	50	⋯	30	90	104	⋯	40	68	16	12	8.1	6.1	53
	575	⋯	50	⋯	30	90	104	⋯	50	86	20	14	10	7.0	67

Figure 16-3 Electrical ratings of ac magnetic contactors and starter. (Courtesy Square D Company.)

Electrical ratings of ac magnetic contactors and starters (NEMA Sizes 4–8):

NEMA Size	Volts	Max HP	Max HP (plugging/jogging †)	Continuous Current (A)	Service-Limit Current (A)	Capacitor kVAR
4	200	40	25	135	156	45
	230	50	30	135	156	52
	380	75	50	135	156	86.7
	460	100	60	135	156	105
	575	100	60	135	156	130
5	200	75	60	270	311	91
	230	100	75	270	311	105
	380	150	125	270	311	173
	460	200	150	270	311	210
	575	200	150	270	311	260
6	200	150	125	540	621	182
	230	200	150	540	621	210
	380	300	250	540	621	342
	460	400	300	540	621	415
	575	400	300	540	621	515
7	230	300		810	932	315
	460	600		810	932	625
	575	600		810	932	775
8	230	450		1215	1400	
	460	900		1215	1400	
	575	900		1215	1400	

Tables and footnotes are taken from NEMA Standards.

† Ratings shown are for applications requiring repeated interruptions of stalled motor current or repeated closing of high transient currents encountered in rapid motor reversal, involving more than five openings or closings per minute and more than ten in a ten-minute period, such as plug-stop, plug-reverse or jogging duty. Ratings apply to single speed and multi-speed controllers.

* Per NEMA Standards paragraph IC 1-21A.20, the service-limit current represents the maximum rms current, in amperes, which the controller may be expected to carry for protracted periods in normal service. At service-limit current ratings, temperature rises may exceed those obtained in normal service. The ultimate trip current of over-current (overload) relays or other motor protective devices shall not exceed the service-limit current ratings of the controller.

★ FLUORESCENT LAMP LOADS — 300 VOLTS AND LESS — The characteristics of fluorescent lamps are such that it is not necessary to derate Class 8502 contactors below their normal continuous current rating. Class 8903 contactors may also be used with fluorescent lamp loads. For controlling tungsten and infrared lamp loads, and resistance heating loads, Class 8903 ac lighting contactors are recommended. These contactors are specifically designed for such loads and are applied at their full rating as listed in the Class 8903 Section.

‡ Ratings apply to contactors which are employed to switch the load at the utilization voltage of the heat producing element with a duty which requires continuous operation of not more than five openings per minute. Class 8903 Types L and S lighting contactors are rated for resistance heating loads.

● When discharged, a capacitor has essentially zero impedance. For repetitive switching by contactor, sufficient impedance should be connected in series to limit inrush current to not more than 6 times the contactor rated continuous current. In many installations, the impedance of connecting conductors may be sufficient for this purpose. When switching to connect additional banks, the banks already on the line may be charged and can supply additional available short-circuit current which should be considered when selecting impedance to limit the current.

The ratings for capacitor switching above assume the following maximum available fault currents:

NEMA Size 2-3: 5,000 A RMS Sym.
NEMA Size 4-5: 10,000 A RMS Sym.
NEMA Size 6-8: 22,000 A RMS Sym.

If available fault current is greater than these values, connect sufficient impedance in series as noted in the previous paragraph.

The motor ratings in the above table are NEMA standard ratings and apply only when the code letter of the motor is the same as or occurs earlier in the alphabet than is shown in the table below.

Motors having code letters occurring later in the alphabet may require a larger controller. Consult local Square D field office.

Motor HP Rating	Maximum Allowable Motor Code Letter
1½-2	L
3-5	K
7½ & above	H

hand, if a fuse were chosen large enough to pass the starting or inrush current, it would not protect the motor against small, harmful overloads that might occur later.

Dual-element or time-delay fuses can provide motor overload protection, but suffer the disadvantages of being nonrenewable and must be replaced.

The overload relay is the heart of motor protection. It has inverse trip-time characteristics, permitting it to hold in during the accelerating period (when inrush current is drawn), yet providing protection on small overloads above the full-load current when the motor is running. Unlike dual-element fuses, overload relays are renewable and can withstand repeated trip and reset cycles without need of replacement. They cannot, however, take the place of overcurrent protective equipment.

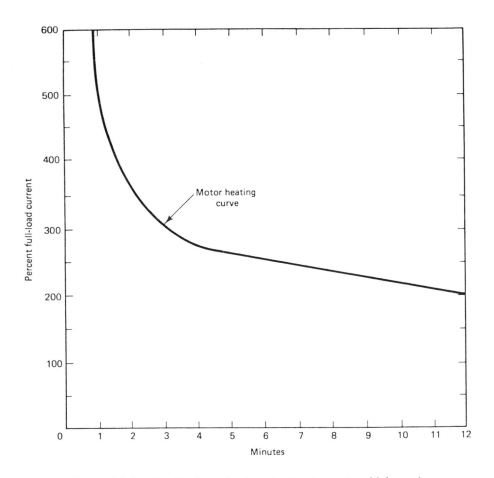

Figure 16-4   The ideal overload protection is one in which sensing properties are very similar to the heating curve of the motor.

The overload relay consists of a current-sensing unit connected in the line to the motor, plus a mechanism, actuated by the sensing unit, that serves to directly or indirectly break the circuit. In a manual starter, an overload trips a mechanical latch and causes the starter contacts to open and disconnect the motor from the line. In magnetic starters, an overload opens a set of contacts within the overload relay itself. These contacts are wired in series with the starter coil in the control circuit of the magnetic starter. Breaking the coil circuit causes the starter contacts to open, disconnecting the motor from the line.

Overload relays can be classified as being either thermal or magnetic. Magnetic overload relays react only to current excesses and are not affected by temperature. As the name implies, thermal overload relays rely on the rising temperatures caused by the overload current to trip the overload mechanism. Thermal overload relays can be further subdivided into two types, melting alloy and bimetallic.

The melting alloy assembly of heater element overload relay and solder pot is shown in Fig. 16-5. Excessive overload motor current passes through the heater element, thereby melting a eutectic alloy solder pot. The ratchet wheel will then be allowed to turn in the molten pool, and a tripping action of the starter control circuit results, stopping the motor. A cooling off period is required to allow the solder pot to "freeze" before the overload relay assembly may be reset and motor service restored.

Melting alloy thermal units are interchangeable and of a one-piece construction, which ensures a constant relationship between the heater element and solder pot and allows factory calibration, making them virtually tamper-proof in the field. These important features are not possible with any other

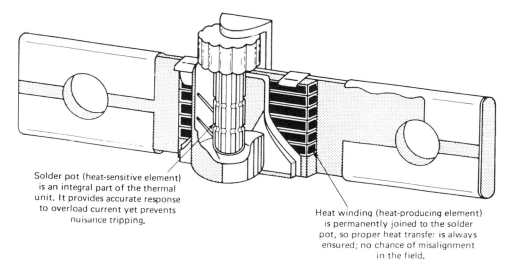

Solder pot (heat-sensitive element) is an integral part of the thermal unit. It provides accurate response to overload current yet prevents nuisance tripping.

Heat winding (heat-producing element) is permanently joined to the solder pot, so proper heat transfer is always ensured; no chance of misalignment in the field.

Figure 16-5   Melting alloy thermal overload relay.

type of overload relay construction. A wide selection of these interchangeable thermal units is available to give exact overload protection of any full-load current to a motor.

Bimetallic overload relays are designed specifically for two general types of application: the automatic reset feature is of decided advantage when devices are mounted in locations not easily accessible for manual operation and, second, these relays can easily be adjusted to trip within a range of 85 to 115 percent of the nominal trip rating of the heater unit. This feature is useful when the recommended heater size might result in unnecessary tripping, while the next larger size would not give adequate protection. Ambient temperatures affect overload relays operating on the principle of heat.

Ambient-compensated bimetallic overload relays were designed for one particular situation, that is, when the motor is at a constant temperature and the controller is located separately in a varying temperature. In this case, if a standard thermal overload relay were used, it would not trip consistently at the same level of motor current if the controller temperature changed. This thermal overload relay is always affected by the surrounding temperature. To compensate for the temperature variations the controller may see, an ambient-compensated overload relay is applied. Its trip point is not affected by temperature and it performs consistently at the same value of current.

Melting alloy and bimetallic overload relays are designed to approximate the heat actually generated in the motor. As the motor temperature increases, so does the temperature of the thermal unit. The motor and relay heating curves (see Fig. 16-6) show this relationship. From this graph, we can see that, no matter how high the current drawn, the overload relay will provide protection, yet the relay will not trip out unnecessarily.

When selecting thermal overload relays, the following must be considered:

1. Motor full-load current
2. Type of motor
3. Difference in ambient temperature between motor and controller.

Motors of the same horsepower and speed do not all have the same full-load current, and the motor nameplate must always be referred to to obtain the full-load amperes for a particular motor. Do not use a published table. Thermal unit selection tables are published on the basis of continuous-duty motors, with 1.15 service factor, operating under normal conditions. The tables are shown in the catalog of manufacturers and also appear on the inside of the door or cover of the motor controller. These selections will properly protect the motor and allow the motor to develop its full horse-power, allowing for the service factor, if the ambient temperature is the same at the motor as at the controller. If the temperatures are not the same, or if the motor service factor is less than 1.15, a special procedure is required to select the proper thermal unit.

Standard overload relay contacts are closed under normal conditions and open when the relay trips. An alarm signal is sometimes required to indicate when a motor has stopped due to an overload trip. Also, with some machines, particularly those associated with continuous processing, it may be required to signal an overload condition, rather than have the motor and process stop automatically. This is done by fitting the overload relay with a set of contacts that close when the relay trips, thus completing the alarm circuit. These contacts are appropriately called *alarm contacts.*

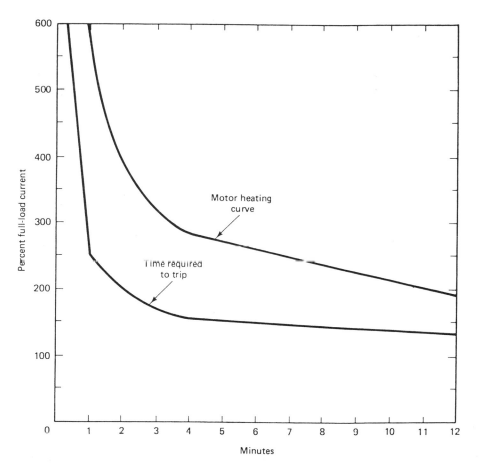

Figure 16-6   Motor and relay heating curves.

A magnetic overload relay has a movable magnetic core inside a coil that carries the motor current. The flux set up inside the coil pulls the core upward. When the core rises far enough, it trips a set of contacts on the top of the relay. The movement of the core is slowed by a piston working in an oil-filled dashpot mounted below the coil. This produces an inverse-time characteristic. The effective tripping current is adjusted by moving the core on a threaded rod. The tripping. time is varied by uncovering oil bypass holes in the piston. Because of the time and current adjustments, the magnetic overload relay is sometimes used to protect motors having long accelerating times or unusual duty cycles.

PROTECTIVE ENCLOSURES

The correct selection and installation of an enclosure for a particular application can contribute considerably to the length of life and trouble-free operation. To shield electrically live parts from accidental contact, some form of enclosure is always necessary. This function is usually fulfilled by a general-purpose sheet steel cabinet. Frequently, however, dust, moisture, or explosive gases make it necessary to employ a special enclosure to protect the motor controller from corrosion or the surrounding equipment from explosion. In selecting and installing control apparatus, it is always necessary to consider carefully the conditions under which the apparatus must operate; there are many applications where a general-purpose enclosure does not afford protection.

The Underwriters Laboratories have defined the requirements for protective enclosures according to the hazardous conditions, and the National Electrical Manufacturers Association has standardized enclosures from these requirements:

NEMA 1: General Purpose

The general-purpose enclosure is intended primarily to prevent accidental contact with the enclosed apparatus. It is suitable for general-purpose applications indoors where it is not exposed to unusual service conditions. A NEMA 1 enclosure serves as protection against dust and light, indirect splashing, but is not dust tight.

NEMA 3: Dust Tight, Rain Tight

This enclosure is intended to provide suitable protection against specified weather hazards. A NEMA 3 enclosure is suitable for application outdoors, on ship docks, canal locks, and construction work, and for application in subways and tunnels. It is also sleet resistant.

NEMA 3R: Rainproof, Sleet Resistant

This enclosure protects against interference in operation of the contained equipment due to rain, and resists damage from exposure to sleet. It is designed with conduit hubs and external mounting, as well as drainage provisions.

NEMA 4: Watertight

A watertight enclosure is designed to meet a hose test that consists of a stream of water from a hose with a 1-inch nozzle, delivering at least 65 gallons per minute. The water is directed on the enclosure from a distance of not less than 10 feet and for a period of 5 minutes. During this period, it may be directed in any direction desired. There shall be no leakage of water into the enclosure under these conditions. Such an enclosure is suitable for application outdoors on ship docks and in dairies, breweries, and the like.

NEMA 4X: Watertight, Corrosion Resistant

These enclosures are generally constructed along the lines of NEMA 4 enclosures except they are made of a material that is highly resistant to corrosion. For this reason, they are ideal in applications such as paper mills, meat packing, and fertilizer and chemical plants where contaminants would ordinarily destroy a steel enclosure over a period of time.

NEMA 7: Hazardous Locations, Class I

These enclosures are designed to meet the application requirements of the NE Code for Class I hazardous locations. In this type of equipment, the circuit interruption occurs in air.

> Class I locations are those in which flammable gases or vapors are or may be present in quantities sufficient to produce explosive or ignitible mixtures.

NEMA 9: Hazardous Locations, Class II

These enclosures are designed to meet the application requirements of the NE Code for Class II hazardous locations.

> Class II locations are those which are hazardous because of the presence of combustible dust.

The letter or letters following the type number indicate the particular group or groups of hazardous locations (as defined in the NE Code) for which the enclosure is designed. The designation is incomplete without a suffix letter or letters.

NEMA 12: Industrial Use

This type of enclosure is designed for use in those industries where it is desired to exclude such materials as dust, lint, fibers and flyings, oil seepage, or coolant seepage. There are no conduit openings or knockouts in the enclosure, and mounting is by means of flanges or mounting feet.

NEMA 13: Oil Tight, Dust Tight

NEMA 13 enclosures are generally made of cast iron, gasketed to permit use in the same environments as NEMA 12 devices. The essential difference is that, due to its cast housing, a conduit entry is provided as an integral part of the NEMA 13 enclosure, and mounting is by means of blind holes, rather than mounting brackets.

NATIONAL ELECTRICAL CODE REQUIREMENTS

The NE Code deals with the installation of equipment and is primarily concerned with safety—the prevention of injury and fire hazard to persons and property arising from the use of electricity. It is adopted on a local basis, sometimes incorporating minor changes or interpretations, as the need arises. NE Code rules and provisions are enforced by governmental bodies exercising legal jurisdiction over electrical installations and are used by insurance inspectors. Minimum safety standards are thus assured.

Motor control equipment is designed to meet the provisions of the NE Code. Code sections applying to industrial control devices are Article 430 on motors and motor controllers and Article 500 on hazardous locations.

With minor exceptions, the NE Code, along with some local codes, requires a disconnect means for every motor. A combination starter consists of an across-the-line starter and a disconnect means wired together in a common enclosure. Combination starters include a blade-disconnect switch, either fusible or nonfusible, while some combination starters include a thermal-magnetic trip circuit breaker. The starter may be controlled remotely with push buttons, selector switches, and the like, or these devices may be installed in the cover. The single device makes a neat as well as compact electrical installation that takes little mounting space.

A combination starter provides safety for the operator, because the cover of the enclosing case is interlocked with the external operating handle of the disconnecting means. The door cannot be opened with the disconnecting means closed. With the disconnect means open, access to all parts may be had, but much less hazard is involved inasmuch as there are no readily accessible parts connected to the power line. This safety feature cannot be obtained with separately enclosed starters. In addition, the cabinet is provided with a means for padlocking the disconnect in the OFF position.

CONTROL CIRCUITS

Two-Wire Control

Figure 16-7 shows wiring diagrams for a two-wire control circuit. The control itself could be a thermostat, float switch, limit switch, or other maintained contact device to the magnetic starter. When the contacts of the control device close, they complete the coil circuit of the starter, causing it to pick up and connect the motor to the lines. When the control device contacts open, the starter is de-energized, stopping the motor.

Two-wire control provides low-voltage release, but not low-voltage protection. When wired as illustrated, the starter will function automatically in response to the direction of the control device without the attention of an operator. In this type of connection, a holding circuit interlock is not necessary.

Three-Wire Control

A three-wire control circuit uses momentary contact, start-stop buttons, and a holding circuit interlock, wired in parallel with the start button to maintain the circuit. Pressing the normally open (NO) start button completes the circuit to the coil. The power circuit contacts in lines 1, 2, and 3 close, completing the circuit to the motor, and the holding circuit contact also closes. Once the starter has picked up, the start button can be released, as the now closed interlock contact provides an alternate current path around the reopened start contact.

Pressing the normally closed (NC) stop button will open the circuit to the coil, causing the starter to drop out. An overload condition, which causes the overload contact to open, a power failure, or a drop in voltage to less than the seal-in value would also de-energize the starter. When the starter drops out, the interlock contact reopens, and both current paths to the coil through the start button and the interlock are now open.

Since three wires from the push-button station are connected into the starter at points 1, 2, and 3, this wiring scheme is commonly referred to as three-wire control (see Fig. 16-8).

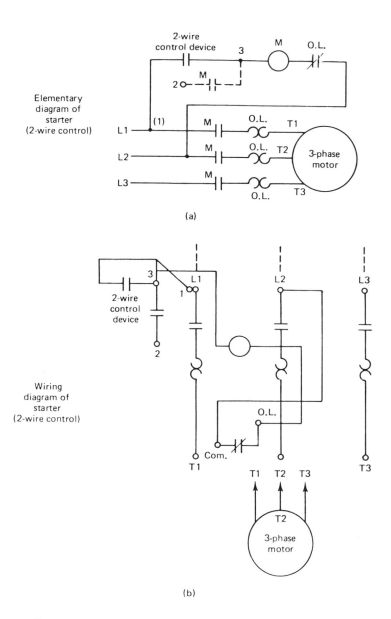

(a)

(b)

Figure 16-7   Elementary diagram of two-wire motor control.
(Courtesy Square D Company.)

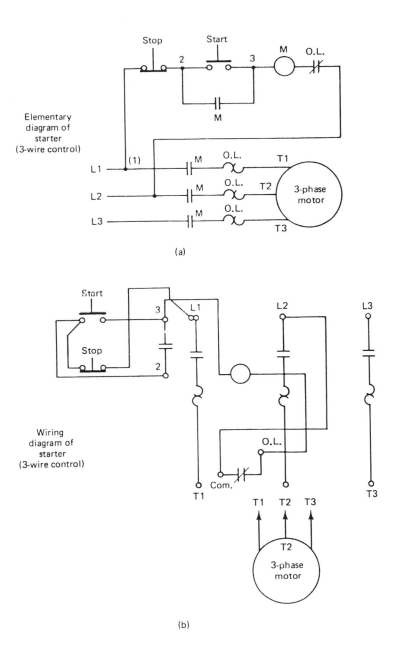

(a)

(b)

Figure 16-8   Elementary diagram of three-wire motor control.
(Courtesy Square D Company.)

The holding circuit interlock is a normally open (NO) auxiliary contact provided on standard magnetic starters and contactors. It closes when the coil is energized to form a holding circuit for the starter after the START button has been released.

In addition to the main or power contacts that carry the motor current, and the holding circuit interlock, a starter can be provided with externally attached auxiliary contacts, commonly called *electrical interlocks.* Interlocks are rated to carry only control-circuit currents, not motor currents. Both NO and NC versions are available. Among a wide variety of applications, interlocks can be used to control other magnetic devices where sequence operation is desired, to electrically prevent another controller from being energized at the same time, and to make and break circuits to indicating or alarm devices such as pilot light, bells, or other signals.

The circuit in Fig. 16-9 shows a three-pole reversing starter used to control a three-phase motor. Three-phase squirrel-cage motors can be reversed by reconnecting any two of the three line connections to the motor. By interwiring two contactors, an electromagnetic method of making the reconnection can be obtained.

**REVERSING MANUAL MOTOR
STARTING SWITCH**

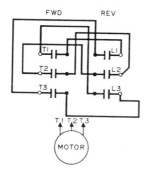

Type K, 3 Pole, 3 Phase

Figure 16-9 Diagram of a three-pole reversing starter used to control a three-phase motor. (Courtesy Square D Company.)

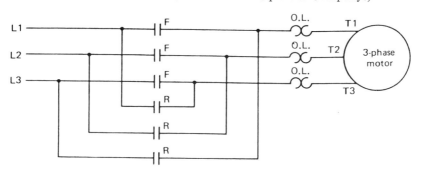

Figure 16-10 Relay amplifying contact capacity. (Courtesy Square D Company.)

As seen in the power circuit (Fig. 16-10), the contacts (F) of the forward contactor, when closed, connect lines 1, 2, and 3 to the motor terminals T1, T2, and T3, respectively. As long as the forward contacts are closed, mechanical and electrical interlocks prevent the reverse contactor from being energized.

When the forward contactor is de-energized, the second contactor can be picked up, closing its contacts (R), which reconnect the lines to the motor. Note that by running through the reverse contacts, line 1 is connected to motor terminal T3, and line 3 is connected to motor terminal T1. The motor will now run in reverse.

Manual reversing starters (employing two manual starters) are also available. As in the magnetic version, the forward and reverse switching mechanisms are mechanically interlocked; but since coils are not used in the manually operated equipment, electrical interlocks are not furnished.

Control Relays

A relay is an electromagnetic device whose contacts are used in control circuits of magnetic starters, contactors, solenoids, timers, and other relays. They are generally used to amplify the contact capability or multiply the switching functions of a pilot device.

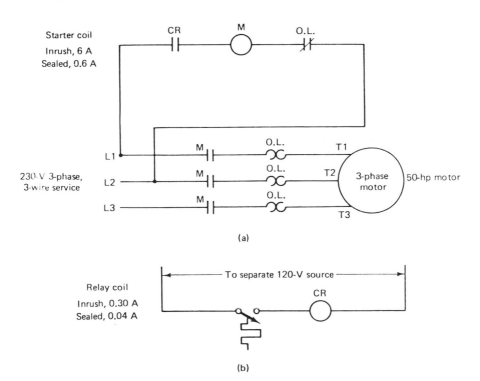

Figure 16-11   Circuit amplifying voltage. (Courtesy Square D Company.)

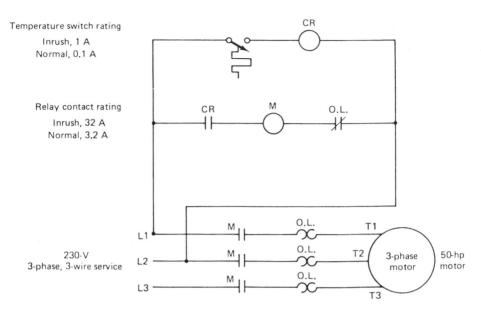

Figure 16-12   (Courtesy Square D Company.)

The wiring diagrams in Fig. 16-11 demonstrate how a relay amplifies contact capacity. Figure 16-11a represents a current amplification. Relay and starter coil voltages are the same, but the ampere rating of the temperature switch is too low to handle the current drawn by the starter coil (M). A relay is interposed between the temperature switch and starter coil. The current drawn by the relay coil (CR) is within the rating of the temperature switch, and the relay contact (CR) has a rating adequate for the current drawn by the starter coil.

Figure 16-11b represents a voltage amplification. A condition may exist in which the voltage rating of the temperature switch is too low to permit its direct use in a starter control circuit operating at some higher voltage. In this application, the coil of the interposing relay and the pilot device are wired to a low-voltage source of power compatible with the rating of the pilot device. The relay contact, with its higher voltage rating, is then used to control the operation of the starter.

Another diagram in Fig. 16-12 represents the use of relays to multiply the switching functions of a pilot device with a single or limited number of contacts. In the circuit shown, a single-pole push-button contact can, through the use of an interposing six-pole relay, control the operation of a number of different loads, such as a pilot light, starter, contactor, solenoid, and timing relay.

Relays are commonly used in complex controllers to provide the logic or "brains" to set up and initiate the proper sequencing and control of a number of interrelated operations. In selecting a relay for a particular application, one of the first steps should be a determination of the control voltage at which the relay will operate. Once the voltage is known, the relays which have the necessary contact rating can be further reviewed, and a selection made, on the basis of the number of contacts and other characteristics needed.

OTHER CONTROLLING EQUIPMENT

Timers and Timing Relays

A pneumatic timer or timing relay is similar to a control relay, except that certain of its contacts are designed to operate at a preset time interval after the coil is energized or de-energized. A delay on energization is also referred to as *on delay*. A time delay on de-energization is also called *off delay*.

A timed function is useful in applications such as the lubricating system of a large machine, in which a small oil pump must deliver lubricant to the bearings of the main motor for a set period of time before the main motor starts.

In pneumatic timers, the timing is accomplished by the transfer of air through a restricted orifice. The amount of restriction is controlled by an adjustable needle valve that permits changes to be made in the timing period.

Drum Switch

A drum switch is a manually operated three-position, three-pole switch that carries a horsepower rating and is used for manual reversing of single- or three-phase motors. Drum switches are available in several sizes and can be spring-return-to-off (momentary contact) or maintained contact. Separate overload protection by manual or magnetic starters must usually be provided, as drum switches do not include this feature.

Push-Button Station

A control station may contain push buttons, selector switches, and pilot lights. Push buttons may be momentary or maintained contact. Selector switches are usually maintained contact, or can be spring return to give momentary-contact operation.

Standard-duty stations will handle the coil currents of contactors up to size 4. Heavy-duty stations have higher contact ratings and provide greater flexibility through a wider variety of operators and interchangeability of units.

Foot Switch

A foot switch is a control device operated by a foot pedal used where the process or machine requires that the operator have both hands free. Foot switches usually have momentary contacts but are available with latches that enable them to be used as maintained contact devices.

Limit Switch

A limit switch is a control device that converts mechanical motion into an electrical control signal. Its main function is to limit movement, usually by opening a control circuit when the limit of travel is reached. Limit switches may be momentary-contact (spring return) or maintained-contact types. Among other applications, limit switches can be used to start, stop, reverse, slow down, speed up, or recycle machine operations.

Snap Switch

Snap switches for motor-control purposes are enclosed, precision switches that require low operating forces and have a high repeat accuracy. They are used as interlocks and as the switch mechanism for control devices such as precision limit switches and pressure switches. They are available also with integral operators for use as compact limit switches, door operated interlocks, and so on. Single-pole, double-throw and two-pole, double-throw versions are available.

Pressure Switch

The control of pumps, air compressors, welding machines, lube systems, and machine tools requires control devices that respond to the pressure of a medium such as water, air, or oil. The control device that does this is a pressure switch. It has a set of contacts operated by the movement of a piston, bellows, or diaphragm against a set of springs. The spring pressure determines the pressures at which the switch closes and opens its contacts.

Float Switch

When a pump motor must be started and stopped according to changes in the water (or other liquid) level in a tank or sump, a float switch is used. This is a control device whose contacts are controlled by the movement of a rod or chain and counterweight, fitted with a float. For closed tank applications, the movement of a float arm is transmitted through a bellows seal to the contact mechanism.

ELECTRONIC CONTROLS

In recent years, many solid-state devices and circuits have entered the motor-control field, reducing previously bulky equipment to compact, efficient, and reliable electronic units. Even electric motors have been built on printed circuit boards with all windings made from flimsy copper foil mounted on a flat card.

The Square D NORPAK^R Solid State Logic Control is one type of control system that has recently been introduced. The NOR is the basic logic element for NORPAK. Using this single element allows many functions to be obtained through the building block approach. While this results in simplicity of design and the minimum of logic elements in a system, it is also a great benefit when changes are required. While NORs can be connected to form any logic function, it is sometimes more convenient to have other logic functions available. For this reason, ANDs, sealed ANDs, ORs, and memories are available. Other functions that cannot be made up from NORs alone are timers, single shots, transfer memories, counters, and shift registers. This variety of components makes this system a wise choice for any application from simple machine control up to complex systems requiring counting and data manipulation. The basic design of a NORPAK system is the same, whether plug-in or encapsulated components are used.

ENERGY MANAGEMENT SYSTEM

The increasing cost of electricity and the shortage of fuel are major items of concern for top management today. Utility bills have risen to levels where action must be taken to help maintain a facility's profitability.

Most commercial and industrial electric bills are made up of two charges, energy and demand. The energy change is based on the quantity of energy consumed for the billing period, while in most cases the demand charge is based on the peak electrical energy used during short periods of time called demand intervals.

One possible solution to lowering these demand and energy costs is through the use of an energy management system, which is a technique of automatically controlling the demand and energy consumption of a facility to a lower and more economical level by shedding and cycling noncritical loads for brief periods of time. This can be accomplished by using demand controllers with their basic features or, where economically feasbiel, by using complex computer-based systems. In some applications, these larger systems are justified, but in many cases smaller, less expensive controllers will do the job.

With the event of the microprocessor, the features of the demand controller along with many of the added features of the computer-based systems have been combined into the Class 8865 EM WATCHDOG Energy Management Systems. The controllers use a continuously integrating demand control technique. It is based on an electronic version of the conventional thermal legged-demand meter. Loads are shed when the predicted demand equals the programmed demand limit.

The priorities, shed, and restore times, along with system data, can be planned out on a data entry worksheet. A calculator-type keyboard is used to enter worksheet data into the controller. A digital readout display confirms that data have been entered.

If a programming error is made, a digital readout will display an error code. To correct an error, the clear entry key is touched and the correct data are reentered. All programming data may then be easily entered or changed.

After all data have been entered, a key switch can be locked into the RUN position, which prevents unauthorized persons from changing critical load data. While in the RUN mode, the controller's digital readout will automatically display the predicted demand, demand limit, and the time of day. All other data settings can be displayed upon request. A battery backup is provided to retain stored information in the event of a power failure.

PROGRAMMABLE CONTROLLERS

Programmable controllers may be used in place of conventional control, such as relays offering faster start-ups, decreased start-up costs, quick program changes, fast troubleshooting, and up-to-date schematic diagram printouts. They also offer additional benefits, such as precision digital timing, counting, data manipulation, remote inputs and outputs, redundant control, supervisory control, process control, management report generation, machine cycle, controller self-diagnostics, dynamic graphic displays, and so forth. Applications include machine tools, sequential, process, conveyor, batching, and energy management control applications.

17

Electric Space Heating

The use of electric heating in residential and commercial buildings has risen tremendously since 1960, and electric heating, at one time, showed all signs of becoming the principal type of heat for all types of structures. However, during the fuel shortage of the 1970s, electricity, along with other types of fuel, rose sharply in price, which made all of us take another look at more economical sources of heat. Still, electric heating has become very popular for use in residential occupancies, mainly because of the following advantages:

1. Electric heat is noncombustible and therefore safer than combustible fuels.
2. It requires no storage space, fuel tanks, or chimneys.
3. It requires little maintenance.
4. The initial installation cost is low.
5. The amount of heat may be easily controlled since each room may be controlled separately with its own thermostat.
6. It is predicted that electricity will be more plentiful than other fuels in the future.

SELECTING HEATING EQUIPMENT

The type of electric heating system used for a given application will usually depend on the structural conditions, the kind of area, and the purpose for which the area will be used. The owner's preference will also enter into the final decision.

Electric heating equipment is available mounted in baseboard (Fig. 17-1a), wall (Fig. 17-1b), ceiling (Fig. 17-1c), kick space (Fig. 17-1d), and floor units; in resistance cable embedded in the ceiling or concrete floor; and in forced-air duct systems similar to conventional oil- or gas-fired furnace hot-air systems except that the source of heat is electric heating elements.

In general, heating equipment should be located on the outside wall near the areas where the greatest heat loss will occur, such as under windows. The controls or wall-mounted thermostats should be located on an interior wall, about 50 inches above the floor, to sense the average temperature.

The baseboard heating units for a residence are laid out as shown in Fig. 17-2. Notice that they are laid out in accordance with the recommendations in the previous paragraph. All these heating units are rated at 240 volts although the wattage ratings of the different units vary.

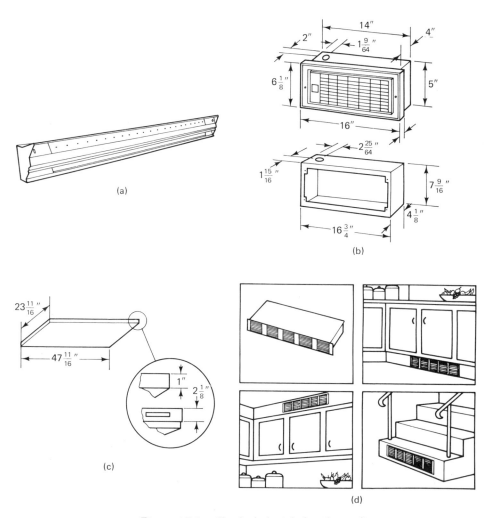

Figure 17-1   Typical electric heating units.

Each branch circuit feeding these units consists of a two-wire cable (armored cable or nonmetallic-sheathed cable with ground wire) run from the panelboard to the units. The cable is size 12 AWG and therefore is protected by a 20-ampere, two-pole circuit breaker in the panel. Since the NE Code requires that circuits feeding electric heating equipment not be loaded to more than 80 percent of their capacity, the maximum allowed wattage that can be connected to each circuit is

$$240 \text{ (volts)} \times 20 \text{ (amperes)} \times 0.80 \text{ (80\%)} = 3840 \text{ watts}$$

The types of electric heat most readily applied to an individually controlled room heating system are as follows:

1. Radiant ceiling heating systems
2. Radiant concrete-floor heating systems
3. Baseboard installations
4. Fan-driven wall heaters

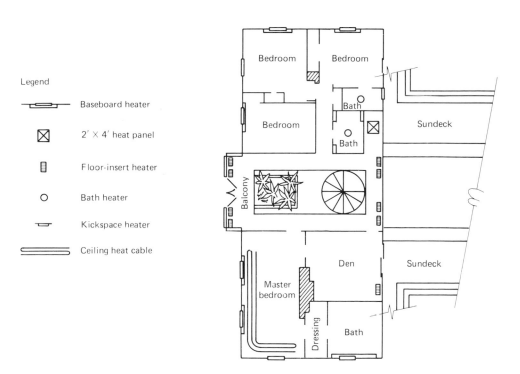

Figure 17-2   Application of electric heating units.

In general, the heating equipment for any room should not exceed the calculated heat loss by more than 10 percent except where limited wattages available from a given manufacturer might prevent this. In that case, equipment most nearly matching the heat loss of the room should be selected.

Ceiling heating cable is manufactured in various lengths and wattages and packaged on reels for field installation. The cable may be stapled to a base consisting of fire-resistant gypsum board or fire-resistant plaster lath. The cable spacing must be in accordance with the NE Code requirement. It is recommended that ceiling cable be installed on 2-inch centers for the first 2 feet of ceiling adjacent to any walls exposed to the outside. Also the cable should be kept at least 8 inches from any junction boxes or outlet boxes and at least 6 inches from any wall. Figure 17-3 gives a pictorial example of heating cable installed in the ceiling of a residence, and Fig. 17-4a shows a typical layout of heating-cable spacing.

It is recommended that the ceiling heating cable be covered with either thermally noninsulating sand plaster, 1/2-inch minimum thickness; plaster manufactured for radiant-heat application; or gypsum board if the cable is first covered with a base coat of plaster or mastic to eliminate air pockets. The manufacturer's specifications for installation of laminated ceiling material should always be followed.

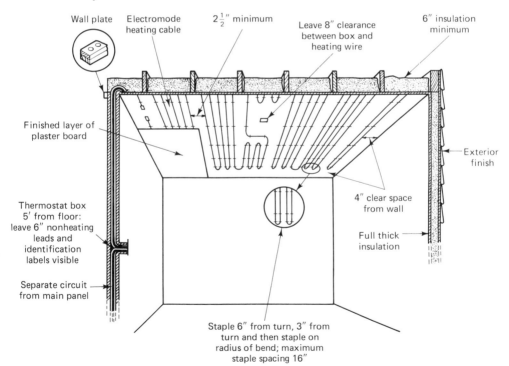

Figure 17-3   Electric heating cable installed in residential ceiling.

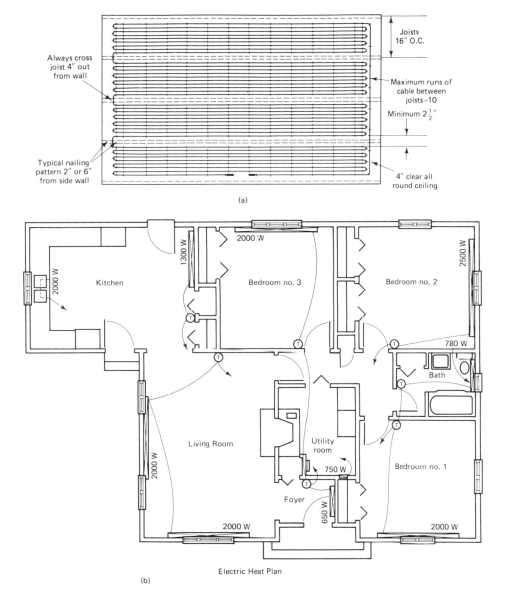

(a)

Joists 16″ O.C.

Always cross joist 4″ out from wall

Maximum runs of cable between joists-10

Minimum $2\frac{1}{2}''$

Typical nailing pattern 2″ or 6″ from side wall

4″ clear all round ceiling

2000 W

1300 W

2000 W

Kitchen

Bedroom no. 3

2500 W

Bedroom no. 2

780 W

Bath

Living Room

Utility room

2000 W

Bedroom no. 1

750 W

Foyer

650 W

2000 W

2000 W

Electric Heat Plan

(b)

Figure 17-4 Example of electric heating cable spacing.

The output wattage of any one heating cable must not exceed 2.75 watts per foot of heating-wire length, nor should the cable be used for drying plaster, as it will result in a change in the plaster composition that will leave a permanent streak in the finished ceiling.

Once the covering has been installed, it may be painted, wallpapered, or covered with any material approved by the cable manufacturer.

Preformed panels are also available. They consist of loose cable embedded in sheets of plasterboard or sheetrock. The preformed panels come complete and ready for installation.

High-intensity radiant heating panels are another type of self-contained heating that can be installed either in the ceiling or the wall. As the name implies, wattage output per foot of service area is higher when compared to other types of systems.

BASEBOARD HEATERS

It is recommended that all baseboard heaters have a metal sheath enclosing the heating element and that the wattage output of any heater not exceed 250 watts per linear foot of unit length. In fact, where wall length permits, it is best that the wattage per linear foot be 200 watts or less. All heaters should also be equipped with temperature-limiting devices that limit the operating temperature of the heating unit to prevent unsafe operation.

Baseboard heaters should be located on the outside wall, if possible, and preferably under windows. However, baseboard units may be located on an inside wall when there are obstructions or available wall space is limited. Figure 17-4b shows the residence with the baseboard heaters and thermostats located and all electrical circuits indicated.

Fan-driven wall heaters must have either a metal-sheath-enclosed element or a nichrome open-wire element. This type of heater should also be equipped with a temperature-limiting device to limit the operating temperature of the heater and thus prevent unsafe operation.

When ceiling heating cable or electric baseboard heaters are used, each room or separate living area should be temperature controlled by its own wall-mounted thermostat (Fig. 17-5). The thermostats should be mounted 48 to 54 inches above the floor and should always be kept away from heat-producing equipment. All line-voltage thermostats should be the two-pole type with an OFF position. Low-voltage thermostats may be used in conjunction with relays.

Fan-driven wall heaters can be controlled by either wall-mounted or built-in thermostats. While electric baseboard heaters may also be controlled with thermostats built into the unit, this practice is not recommended.

S3, S4   S3-T, S4-T

1F56-301

T834C1004

T87F1800 Mounted
on Q539 Subbase

1F86-2

T8082A1015

Figure 17-5   Typical wall thermostats.
(Courtesy Square D Company.)

HUMIDITY CONTROL

For adequate moisture control in the living area of the home, the following are recommended:

1. At least one exhaust fan should be installed. A kitchen-hood fan is usually sufficient.
2. The exhaust fan should not be rated less than 300 cubic feet per minute.
3. The fan should be vented to the outside of the building through the roof or wall.
4. All exhaust fans should be equipped with automatic dampers.
5. Control of the exhaust fan should be accomplished by means of a humidistat connected parallel to the manual fan switch.
6. The humidistat should be set at 35 to 40 percent relative humidity for wintertime operation; this will cause the fan to operate when the humidity exceeds this setting.
7. The humidistat should be set at 100 percent or at OFF during the summer months to prevent continuous fan operation.
8. The duct connecting the exhaust fan with the outside louver should be wrapped or covered with insulation to prevent the warm moist air from condensing in the duct and dropping back into the room.

CIRCUIT REQUIREMENTS

As mentioned previously, when electric heat is supplied by a duct system in the form of electric heating elements in a furnace or ductwork, the work is normally handled by a mechanical contractor. However, the electrical connections (feeders) are handled by the electrical contractor and thus are the responsibility of the electrical designer, too.

Circuit requirements for this type of heating equipment do not greatly differ from those for any other circuit, but the electrical designer must carefully study the manufacturer's electrical wiring diagrams for the equipment in question to determine exactly the number and type of circuits required for the system.

For example, an electric furnace may have a total rating of 24 kilowatts. It would be natural for an electrical designer to calculate the circuit size for the unit as follows:

$$\frac{24,000 \text{ (watts)}}{240 \text{ (volts)}} \times 1.25 \text{ (allowance per code)} = 125 \text{ amperes}$$

Using this calculation as a basis, the electrician normally would use a circuit rated at a minimum of 125 amperes. However, resistance-type heaters found in a furnace of this type usually will be connected in steps, that is, four elements of 6 kilowatts each connected in parallel. If the one circuit is used for the feeder, a subpanel would have to be installed to separately feed each of the four heaters. There would have to be four 40-ampere circuits feeding each heating element. If the designer did not specify this, the electrical contractor would be entitled to an extra billing, costing someone additional money. Sometimes such electric furnaces have their own controls built into the unit, which would separate the 125-ampere circuit into four subcircuits, but this information must be verified with each separate project or piece of heating equipment.

The examples in Figs. 17-6 through 17-9 show several applications of how large electric heaters appear on working drawings. A study of these should aid the reader in determining a suitable method to use for nearly every residential application.

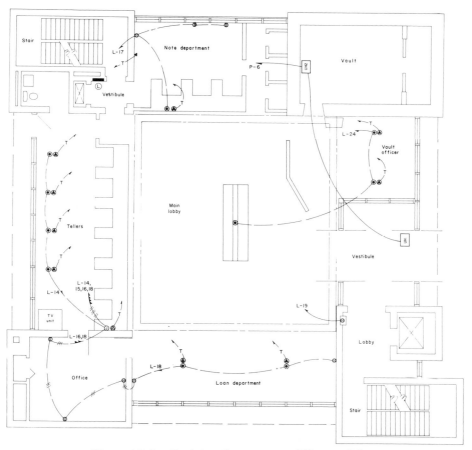

Figure 17-6   Bank heating system.   UH = unit heater.

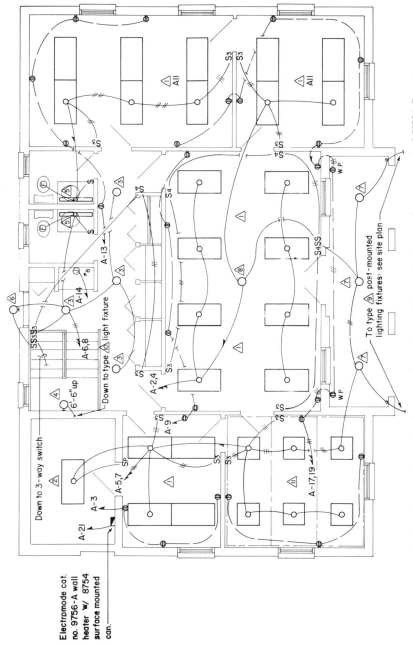

Figure 17-7 Electric fan-forced heater used in vault to supplement central HVAC system.

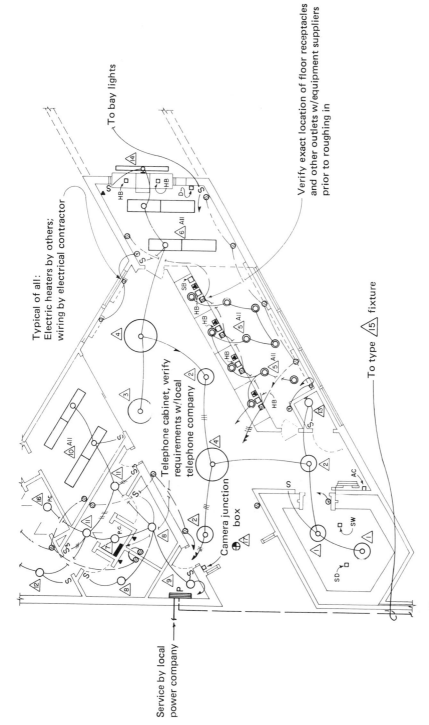

Figure 17-8 Baseboard heaters used in front of windows to supplement central HVAC system.

To bay lights

Verify exact location of floor receptacles and other outlets w/equipment suppliers prior to roughing in

Typical of all:
Electric heaters by others; wiring by electrical contractor

Telephone cabinet, verify requirements w/local telephone company

Camera junction box

Service by local power company

To type 15 fixture

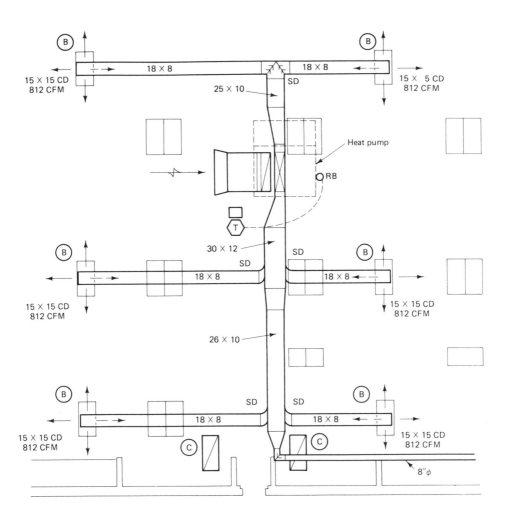

Figure 17-9

T PUMPS

The term *heat pump,* as applied to a year-round air-conditioning system, commonly denotes a system in which refrigeration equipment is used in such a manner that heat is taken from a heat source and transferred to the conditioned space when heating is desired; heat is removed from the space and discharged to a heat sink when cooling and dehumidification are desired. Therefore, the heat pump is essentially a heat-transfer refrigeration device that puts the heat rejected by the refrigeration process to good use. A heat pump can do the following:

1. Provide either heating or cooling.
2. Change from one to the other automatically as needed.
3. Supply both simultaneously if so desired.

A heat pump has the unique ability to furnish more energy than it consumes. This uniqueness is due to the fact that electric energy is required only to move the heat absorbed by the refrigerant. Thus, a heat pump attains a heating efficiency of two or more to one; that is, it will put out an equivalent of 2 or 3 watts of heat for every watt consumed. For this reason its use is highly desirable for the conservation of energy.

Water-to-air heat pumps, as opposed to air-to-air heat pumps, have the following advantages:

1. The pumps can be located anywhere in the building since no outside air is connected to them.
2. The outside air temperature does not affect the performance of the heat pump, as it does an air-to-air type of heat pump.
3. Since the water source of the water-to-air heat pump will seldom vary more than a few degrees in temperature, a more consistent performance can be expected.

The schematic drawings in Fig. 17-10 show how two heat pumps were connected for a residence; the swimming-pool water was used as a means of heat exchange, and since this water was preheated to approximately 78°F, the efficiency of the heat pumps utilizing this same water approached the maximum.

During the warm months the heat pumps were reversed (cooling cycle) to cool the area. Again, the pool water was used as the heat exchange, this time acting as a "cooling tower." The air from these heat pumps was

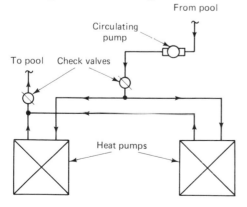

Figure 17-10

distributed by means of underfloor transit ducts with supply-air diffusers and return-air grilles raised above the floor level and mounted in a wainscot around the perimeter of the area. This was done to prevent water from entering the openings in the ductwork.

HEATING CALCULATIONS

The following are the goals of an electric heat installation:

1. Adequate, dependable, and trouble-free installation.
2. Year-round comfort.
3. Reasonable annual operating cost.
4. Reasonable installation cost.
5. Systems that are easy to service and maintain.

Heat-loss calculations must be made to ensure that heating equipment of proper capacity will be selected and installed. Heat loss is expressed in either Btu's per hour (which is abbreviated Btuh) or in watts. Both are measures of the rate at which heat is transferred and are easily converted from one to the other:

$$\text{Watts} = \frac{\text{Btuh}}{3.4}$$

$$\text{Btuh} = \text{watts} \times 3.4$$

Basically, the calculation of heat loss through walls, roof, ceilings, windows, and floors requires three simple steps:

1. Determine the net area in square feet.
2. Find the proper *heat-loss factor* from a table.
3. Multiply the area by the factor; the product will be expressed in Btuh. Since most electric heat equipment is rated in watts rather than Btuh, divide this product by 3.4 to convert to watts.

Calculations of heat loss for any building may be made more quickly and more efficiently by using a prepared form, such as that shown in Fig. 17-11. With spaces provided for all necessary data and calculations, the procedure becomes routine and simple.

Load estimate is based on design conditions inside the building and outside in the atmosphere surrounding the building. Outside design conditions are the maximum extremes of temperature occurring in a specific locality. The inside design condition is the degree of temperature and humidity that will give optimum comfort.

HEATING LOAD				
1. DESIGN CONDITIONS		DRY BULB (F)	SPECIFIC HUMIDITY gr./lb.	
OUTSIDE		$0° F$		
INSIDE		$70° F$		
DIFFERENCE		$70° F$		
	ITEMS		HEAT LOAD (BTUH)	

2. TRANSMISSION GAIN (TABLE 2)						
WINDOWS	SQ. FT.	X	FACTOR	X	DRY BULB TEMP. DIFF.	
						=
WALLS						=
						=
						=
						=
						=
ROOF	5 4		0.09		70	= 340.2
FLOOR	5 4		0.20		70	= 756.0
OTHER						=

3. VENTILATION OR INFILTRATION (TABLES 5 AND 6, USE LARGER QUANTITY)						
SENSIBLE LOAD	CFM	X	DRY BULB TEMP. DIFF.	X	FACTOR	
	7.6		70		1.08	= 574.6
HUMIDIFICATION LOAD	CFM		SPECIFIC HUMIDITY DIFF.	X		
					0.67	=

4. DUCT HEAT LOSS (TABLE 7)	
HEAT LOSS _____ X FACTOR FOR INSULATION THICKNESS _____ X DUCT LENGTH (ft.) _____ ÷ 100	=
5. TOTAL HEATING LOAD	=

Figure 17-11

18

Air Conditioning

While the installation of air-conditioning systems usually falls under the mechanical section of the building's systems, electricians are required to furnish power, and sometimes do the control wiring, for such systems. On smaller projects, such as residential construction and small commercial projects, the electrical contractor may be called on to furnish and install such items as ventilating fans, through-wall room air conditioners, duct heaters, and the like. For these reasons, anyone involved in electrical technology should have a basic understanding of air-conditioning systems.

VENTILATION AND CIRCULATION

Ventilation is the moving of air from one space to an entirely separate space and is primarily a matter of air volume. *Circulation* is the moving of air around within a confined space and is primarily a matter of air velocity.

Both ventilation and circulation utilize fans and are subject to the same basic laws of physics, which consist mainly of the following factors when selecting or specifying any fan:

1. Volume of air required in cubic feet per minute (CFM)
2. Static pressure (SP)
3. Type of application
4. Maximum tolerable noise level

5. Ambient and airstream temperature
6. Belt or direct drive
7. Available line voltage, phase, and frequency

Very basic and simple principles of fluid dynamics govern the performance of fans and their relationship to the ventilation system. There is no need for rigorous engineering understanding of these principles, however, in the usual day-to-day installation of such systems or in the simpler ventilation problems most frequently encountered. However, you should have knowledge of the following terms.

Air horsepower (AHP): Work done in moving a given volume or weight of air at a given speed.

Area (A): The square feet of any plane surface or cross section of a duct, the air inlet or outlet from a room, or the circular plane of rotation of a propeller.

Barometric pressure: Pressure expressed by the height in inches of a column of mercury, exerted by the weight of the earth's atmosphere on any surface.

Brake horsepower (BHP): Work done in driving any fan. This load, plus drive losses from belts and pulleys, is the work done by a fan's electric motor. It is always a higher number than AHP.

Cubic feet per minute (CFM): The physical volume (not weight) of air moved by a fan.

Density: The actual weight of air in pounds per cubic foot.

Mechanical efficiency (ME): A decimal number or a percentage representing the ratio between air horsepower divided by brake horsepower of a fan, which is always less than 1.000 or 100 percent.

Pressure: Any fan produces a total pressure. This is the sum of the velocity pressure, which is always positive, and the static pressure, which is usually positive on the outlet side and negative on the inlet side of any fan. Thus, total pressure equals velocity pressure plus static pressure. Velocity pressure results only when air is in motion. Static pressure, in its general effect, may be likened to friction and may be described as the pressure that tends to explode a duct (if positive) or collapse it (if negative). All these pressures are expressed in inches of a column of water that they will support.

Revolution per minute (rpm): The speed at which a fan or motor turns (revolves).

Standard air: Most fan rating charts or curves are shown at standard air to provide a uniform basis for comparison. By definition, standard air has a density of 0.075 pounds per cubic foot, which is the weight of 1 cubic foot of dry air at a temperature of 70°F and a barometric pressure of 29.92 inches of mercury. Outside of laboratory-controlled conditions, standard air seldom exists in fan work, but the existing environment is usually close enough to standard for practical purposes.

Temperature: The dry-bulb temperature of either ambient air or exhaust-stream air. Most fans will work satisfactorily at temperatures up to about 104°F. If the temperature is higher, calculations must be made to determine what effect the temperature will have on satisfactory operation.

Velocity: The speed in feet per minute at which air is moving at any location, such as through a duct, inlet damper, outlet shutter, or at the fan discharge point.

FAN EQUATIONS

The following equations will produce the facts needed for fan application work:

1. A (Area) $\times$ V (velocity) = CFM
2. CFM/V = A
3. CFM/A = V

The basic fan laws are employed in calculating fan performance. They rely on the fact that the efficiency (ME) of a fan remains constant throughout its useful range of operating speeds (fan rpm). When fan speed is varied,

4. Air delivery will vary directly with the rpm ratio:

$$\text{New CFM} = \frac{\text{new rpm}}{\text{old rpm}} \times \text{old CFM}$$

5. Fan (and system) pressure will vary directly with the square of the rpm ratio:

$$\text{New SP} = \left(\frac{\text{new rpm}}{\text{old rpm}}\right)^2 \times \text{old SP} \quad \text{(or TP or VP)}$$

6. Brake horsepower load on the fan motor (or air horsepower on the fan) will vary directly with the cube of the rpm ratio:

$$\text{New BHP} \quad \text{(or AHP)} - \left(\frac{\text{new rpm}}{\text{old rpm}}\right)^2 \times \text{old BHP} \quad \text{(or AHP)}$$

The preceding are by no means all the principles and equations associated with fan and ventilation system engineering, but they are the ones that are the most useful in practical everyday applications. Figure 18-1 shows a graphic illustration of the fan laws. This graph illustrates typical propeller fan performance, where BHP rises as pressure rises. Centrifugal fan curves are similar, but BHP drops as pressure rises.

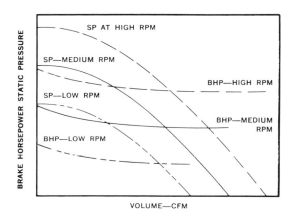

Figure 18-1   Typical propeller fan performance where BHP rises as pressure rises. Centrifugal fan curves are similar, but BHP drops as pressure rises. (Courtesy Airmaster Fan Co.)

SERIES AND PARALLEL FAN OPERATION

Two fans operated in series do not double the CFM. Instead, total airflow is limited to the CFM capacity of one fan alone. Series operation is seldom desirable unless it is necessary to maintain constant pressure, zero pressure, or a constant vacuum in the room.

Fans operated in parallel produce a total airflow equal to the sum of their individual CFM capacities. Parallel operation is desirable from an airflow-distribution standpoint. It is also frequently the only way to move the total volume of air required where space limitations prohibit the use of one larger fan.

GENERAL VENTILATION

Moving a volume of air through a large space, where there need be no concern about a concentrated source of heat or contamination, is called *general ventilation.* Usually no particular velocity is required but rather a specific volume of air. The two general methods of determining the needed CFM are discussed next.

Determining CFM by the Air-Change Method

First determine the total cubic feet of air space in the building. Then consult the table to find the required number of air changes necessary to give proper ventilation.

$$\text{CFM} = \frac{\text{building volume in cubic feet}}{\text{minutes per air change}}$$

To give an example, assume the dimensions of a building to be ventilated are 100 by 30 feet with a 15-foot ceiling. Multiply 100 × 30 × 15, which gives 45,000 cubic feet of air space. If a 3-minute air change is required, 45,000 cubic feet divided by 3 equals 15,000 CFM needed to change the air every 3 minutes (or 20 air changes every hour).

Average air changes recommended for good ventilation in various areas are shown in Fig. 18-2. Comfort cooling, which is the rate of air change fast enough to provide good ventilation as well as comfort by effectively removing heat, is obtained by using the lower figure for each range shown in the table in Fig. 18-2. Several factors will affect the selection within the stated ranges. Geographic locations enter the calculations. The warmer the climate is, the more air required for good ventilation. Increased ventilation is required where large groups of people work in a small area. Also, ceiling height is a factor. For ceilings 8 to 15 feet high, the air change in minutes will vary proportionally, with more ventilation being required for the 8-foot ceiling.

Once these calculations are made, refer to a manufacturer's catalog or specifications to select the proper fan that will deliver the total CFM required. Existing conditions will also affect the installation. All dead air space should be eliminated through proper fan location or use of supplementary air circulator fans. Install the fan on the lee side of the building whenever possible to take advantage of prevailing winds. It should also be installed opposite the intake area so that air moves the entire length of the room to be ventilated.

Determining CFM by the Heat-Removal Method

In certain installations where ventilation of heat is required, such as in refrigeration and manufacturing areas, the BTU per minute is first established, followed by the average outside temperature and the desired inside temperature. The following equation gives the amount of air to be passed through the building to maintain desired inside temperature:

$$\text{CFM} = \frac{\text{total BTU per minute}}{0.0175 \times \text{temperature, rise} \,^\circ\text{F}}$$

In this equation, the CFM deals primarily with sensible heat, which is heat that changes the temperature of the substance involved. Radiant heat is not considered.

	Minutes per Change
Assembly halls	2–10
Auditoriums	2–10
Bakeries	2–3
Banks	3–10
Barns	10–20
Bars	2–5
Beauty parlors	2–5
Boiler rooms	1–5
Bowling alleys	2–10
Churches	5–10
Clubs	2–10
Dairies	2–5
Dance halls	2–10
Dining rooms	3–10
Dry cleaners	1–5
Engine rooms	1–3
Factories	2–5
Forge shops	2–5
Foundries	1–5
Garages	2–10
Generator rooms	2–5
Gymnasiums	2–10
Kitchens, hospitals	2–5
Kitchens, residential	2–5
Kitchens, restaurant	1–3
Laboratories	1–5
Laundries	1–3
Markets	2–10
Offices	2–10
Packing houses	2–5
Plating rooms	1–5
Pool rooms	2–5
Projection rooms	1–3
Recreation rooms	2–10
Residences	2–5
Sales rooms	2–10
Theaters	2–8
Toilets	2–5
Transformer rooms	1–5
Warehouses	2–10

Figure 18-2   Average air changes required for good ventilation. (Courtesy Airmaster Fan Co.)

Make-up air is necessary for all exhaust ventilation because what goes out must first come in. Therefore, adequate intake area must be provided when fans are used to exhaust air or adequate exhaust area when fans supply intake air. The same general rules apply to both cases.

The velocity of air intake should not be too high if ingestion of rain, snow, or airborne debris will be harmful or undesirable. Generally, an intake velocity of about 400 feet per minute (FPM) through a louver or damper will exclude weather. Through windows, doors, or other open intake areas, 300 to 500 FPM usually works. However, always consider local conditions in making a choice.

Doors and windows are suitable intake areas if they are located close enough to the floor and provide a full sweep through the area to be ventilated. Otherwise, intake louvers or adjustable dampers should be provided. Determine the required intake area as follows:

$$\text{Square feet free intake area} = \frac{\text{CFM}}{\text{FPM}}$$

For example, assume that a room requires 30,000 CFM with a 300-FPM velocity. The volume of air (30,000 CFM) divided by 300 results in 100 square feet of free intake required. Deduct area for windows and doors. If more is needed, fixed or adjustable louvers may be needed to regulate the amount of air intake because of winter conditions or other reasons.

SCREEN EFFICIENCY

Many locations necessitate the use of bird or insect screen over intake areas. Such screen reduces the free intake area and necessitates a larger overall area for compensation. The sketches in Fig. 18-3 illustrate the effective free area of three types of screen. Choose the appropriate one and multiply the free intake area by the percent of efficiency to obtain the size needed.

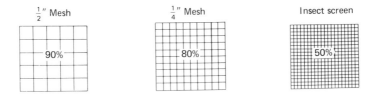

Figure 18-3   Examples of different-sized screen mesh.
(Courtesy Airmaster Fan Co.)

STATIC PRESSURE

In general ventilation, static pressure is usually very low provided intake area is adequate. Intake louvers or dampers and exhaust shutter do impose some losses, and static pressure of 1/8 inch should be assumed when they are used. If windows and doors are used for intake, little static pressure exists when they are open. However, during colder weather when doors and windows are closed, static pressure can rise, sometimes to as much as 1/4 inch. Be sure to consider all these conditions when selecting a fan.

CHECKLIST FOR PROPER FAN INSTALLATION

1. Install fans and intake openings at opposite ends of the enclosure so that intake air will sweep lengthwise through the area to be ventilated.
2. The net intake area must be at least 30 percent greater than the exhaust fan orifice.
3. Fans should blow with prevailing winds. Install on the lee side.
4. Intake areas should be located on the windward side to utilize pressure produced by prevailing winds.
5. If extreme quiet is necessary, fans should be spring mounted and connected to the wall opening by a canvas boot.
6. If steam, heat, or odors are to be exhausted, mount the fans near the ceiling, use totally enclosed motors, and locate intake areas near the floor.
7. If exhaust air is hazardous, use an explosion-proof motor or mount the motor outside the airstream. Also specify a spark-proof fan.
8. Fans exhausting air in close back-to-back proximity should be avoided. If unavoidable, separate them by three or more fan diameters.
9. Where filters are used, intake area must be increased. Get the filter manufacturer's exact resistance figures or increase the intake area three or four times to allow minimum pressure loss from resistance of the filter.

CHECKLIST FOR POSSIBLE ERRORS IN FAN INSTALLATION

1. Be careful not to select a fan of too low capacity for the job.
2. Don't use one large fan instead of two smaller ones. Two will usually provide better air distribution and more operating flexibility.
3. Don't use propeller fans on long duct runs unless they are designed to operate against static pressure.

4. Don't try to force air through ducts smaller than the area of the fan.
5. Don't increase fan rpm to increase CFM output. It may cause motor burnout.
6. Don't use insect screen with exhaust fans. It reduces efficiency, requires frequent cleaning, and can cause motor burnout.
7. Where two shutters are used on one fan installation, the intake shutter should be motor operated.
8. Fans that exhaust air from enclosed indoor rooms require adequate intake area into the room.
9. Standard fans should not be exposed to abrasive or corrosive conditions without proper treatment.

LOCAL VENTILATION

Localized ventilation is used to control atmospheric contamination or excessive heat *at its source* with a minimum of airflow and power consumption. Air velocity past the contaminant source must be fast enough to capture grease, fumes, paint spray, or other materials and carry them into an exhaust hood. For example, a paint spray booth must be designed to draw from 100 to 175 FPM capture velocity across the open-face area of the booth. Velocity at the discharge duct must be 1400 to 2000 FPM. Recommended capture and duct velocity tables are shown in Fig. 18-4. Recommended velocity of air for exhaust systems is shown in Fig. 18-5.

The open-face area of any booth is determined by its physical size and the required access to the work area. It may be a simple open-front booth, or additional openings may be needed on the sides, as in the case of conveyorized transport of work, or it may be an island hood open on four sides. All open-face areas must be added together. Then the fan CFM capacity required is easily determined by the following equation:

$$\text{CFM} = \text{face area (ft}^2) \times \text{face velocity (FPM)}$$

The static pressure (SP) at which the fan must work can often be approximated without going to the trouble of making detailed calculations if these eight simple steps are followed:

1. Assume the losses in the hood itself to be 0.05- to 0.10-inch SP.
2. Size the cross-sectional area of the duct to ensure 1400 to 2000 FPM velocity.

$$\text{Duct area (ft}^2) = \frac{\text{CFM}}{\text{FPM}}$$

3. Keep all ductwork as short and straight as possible. Avoid elbows and sharp turns.

Process	Type of Hood	Capture Velocity (FPM)
Aluminum furnace	Enclosed hood open one side	150-200
	Canopy or island hood	200-250
Brass furnace	Enclosed hood open one side	200-250
	Canopy hood	250-300
Chemical laboratory	Enclosed hood, front opening	100-150
Degreasing	Canopy hood	150
	Slotted sides, 2"-4" slots	1500-2000
Electric welding	Open front booth	100-150
	Portable hood, open face	200-250
Foundry shake-out	Open front booth	150-200
Kitchen ranges	Canopy hoods	125-150
Paint spraying	Open front booth	100-175
Paper drying machine	Canopy hood	250-300
Pickling tanks	Canopy hood	200-250
Plating tanks	Canopy hood	225-250
	Slotted sides	250 CFM per ft. of tank surface
Steam tanks	Canopy hood	125-175
Soldering booth	Enclosed booth, open one side	150-200

Figure 18-4 Capture velocity of various types of booths.
(Courtesy Airmaster Fan Co.)

	Quietness Important (FPM)	Quietness Not Important (FPM)
Main ducts	1000 to 1500	1500 to 2200
Branch ducts	800 to 1200	1000 to 1500
Grilles	500	800

Figure 18-5 Recommended velocity of air from exhaust systems. (Courtesy Airmaster Fan Co.)

4. Determine the total straight duct length.
5. Add the equivalent straight duct length for each turn from the chart in Fig. 18-6 and add it to step 4.
6. Multiply the total length by 0.0025 inch to determine an approximate SP for the ductwork.
7. For booths with washable or replaceable filters, assume 0.25-inch SP for clean or 0.50-inch for dirty filters, or obtain the filter manufacturer's recommendations.
8. Add the SP figures from steps 1, 6, and 7 to obtain the approximate total SP.

You now have the CFM and SP required for sizing the fan from manufacturer's data. Figures 18-7, 18-8, and 18-9 show examples of ventilating systems.

Dia. of Pipe	90° Elbow Centerline Radius		
	1.5D	2.0D	2.5D
3"	5	3	3
4"	6	4	4
5"	9	6	5
6"	12	7	6
7"	13	9	7
8"	15	10	8
10"	20	14	11
12"	25	17	14
14"	30	21	17
16"	36	24	20
18"	41	28	23
20"	46	32	26
24"	57	40	32
30"	74	51	41
36"	93	64	52
40"	105	72	59
48"	130	89	73

For 60° elbows — x.67
For 45° elbows — x.5

Figure 18-6   Equivalent resistance in feet of straight pipe.

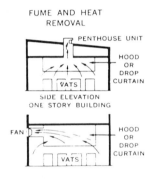

FUME AND HEAT REMOVAL

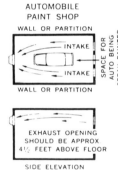

AUTOMOBILE PAINT SHOP

Figure 18-7   Examples of ventilation systems.

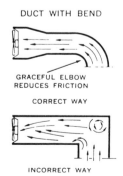

DUCT WITH BEND

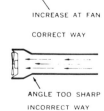

STRAIGHT DUCT

Figure 18-8   Various duct designs. (Courtesy Airmaster Fan Co.)

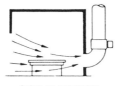

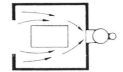

SIDE ELEVATION

EXHAUST OPENING SHOULD BE
APPROX 4½ FEET ABOVE FLOOR

PLAN VIEW

Figure 18-9 Plan and elevation of
paint spray booth ventilation.
(Courtesy Airmaster Fan Co.)

AIR CIRCULATORS

There are four main uses for fans that circulate air through an enclosed space at appreciable velocity:

1. To create a curtain of high-velocity air to prevent the movement of insects or the convective movement of air between adjoining spaces.
2. To eliminate dead air spots within an enclosure.
3. To cool hot materials, machinery, or parts by moving cooler air across their surfaces.
4. To cool people. This is by far the most common use for air-circulating fans.

In the effective cooling of people, air movement utilizes favorable phenomena in human physiology. Human bodies have a built-in temperature-regulation system. Their metabolism produces body heat. To maintain thermal equilibrium, the body must lose heat at exactly the same rate as its metabolism produces it.

The body rejects excess heat by evaporation, convection, and radiation, all of which are affected by environment. Only the first two can be assisted by circulating fans. Fans will not materially affect body heat loss or gain resulting from radiation. A man working near a red-hot furnace can be better protected by installation of a shiny aluminum shield between his position and the infrared source.

Fans will help the body lose heat by convection provided the temperature of the air is lower than the surface temperature of the skin. The convection loss is limited by the temperature differential, as well as by human foibles.

Of substantial benefit is the assistance fans give to body heat losses by evaporation. It is, in effect, air conditioning. When perspiration changes from the solid to the gaseous state for any reason, heat is absorbed by the vapor. If perspiration can be induced to evaporate at a faster rate, more heat is removed. This phenomenon is called *evaporative cooling.*

Any movement of air across a sweating surface increases cooling. The higher the velocity of air movement, the higher the rate of comfort cooling up to the limits of human tolerance. As a broad rule of thumb, general ventilation fans are primarily an aid to convective heat losses. But air circulators help where it means the most, when the temperature is high, by accelerating evaporative cooling.

LOW-POWERED FANS

Low-powered fans can be defined as fans that in most cases have small fractional-horsepower motors, usually 1 horsepower or less. They are usually axial-flow propeller fans, often without weather-protection housings. Some, intended for roof-top mounting, do have weather covers.

There are many ways to conserve energy and save money by the proper application of low-powered fans. The most common applications almost always fall into some combination of the following categories:

1. Minimizing heat losses
2. Reclaiming waste heat
3. Reducing electric power consumption
4. Preventing damage to materials and structures
5. Increasing bodily comfort

Many of the most useful applications are in homes, but commercial and industrial applications are almost as common and are equally effective. For the most part, these are projects of the do-it-yourself type. Bigger, higher-cost projects are usually engineered and installed by experts.

The following are suggestions for use of low-powered fans to conserve energy:

1. Insulate the building.
2. Select the best type of fan for the job.
3. Select the most efficient fan available.
4. Use the best installation techniques possible.

19

Planning a Residential Wiring Installation

With the exception of very large residences and tract-development houses, the size of the average residential electrical system has not been large enough to justify the expense of preparing complete electrical working drawings and specifications. Usually such electrical systems are laid out by the architect or owners in the form of a sketchy outlet layout, requiring that the electricians installing the systems lay out and design the system as the work progresses. Therefore, those involved in residential wiring must have a good knowledge of the NE Code, wiring methods, and the like, in order to perform this work correctly and efficiently.

The primary use of early home electrical systems was to provide interior lighting, but today's uses of electricity include much more:

1. Heating and air conditioning
2. Electrical appliances
3. Electrical cooking
4. Interior and exterior lighting
5. Communication systems
6. Alarm systems

GENERAL PLANNING

In planning any electrical system, certain general factors must be considered regardless of the type of construction.

1. Wiring method
2. Overhead or underground electrical service
3. Type of building construction
4. Type of service entrance and equipment
5. Grade of wiring devices and lighting fixtures
6. Selection of lighting fixtures
7. Type of heating and cooling system
8. Control wiring for heating and cooling
9. Signal and alarm systems

The experienced technician readily recognizes, within certain limits, the type of system that will be required. However, always check the local code requirements when selecting a wiring method. If more than one wiring method will be utilized, decide which type of system will be provided prior to the preparation of the wiring layout.

In a residential occupancy, a 120/240 volt, single-phase service will invariably be provided by the local utility company. Therefore, the feeders will be three-wire, and the branch circuits will be either two- or three-wire cables. The safety switches, service equipment, and panelboards will be three-wire solid neutral. On each project, however, it must be determined where the point of service-drop attachment to the building will be located and whether the service is to be provided as part of the contract or by the utility company.

LIGHTING OUTLETS

Lighting fixtures should be carefully selected and located to afford the best distribution of light; yet they should fit in with the architectural scheme of the structure and the interior decor. Often a lighting fixture allowance will be made by the contractor and then the owners make the selection within the alloted amount. However, usually the best method is to let those who are familiar with residential illumination make the selection and then meet with the owners to discuss the fixtures selected. This will almost always result in a better lighting design.

CONVENIENCE OUTLETS

Convenience outlets are usually positioned at presumed convenient locations to provide electrical power for portable appliances. Their location must be in accordance with the NE Code and also meet the requirements of being convenient; otherwise, they might be "inconvenient outlets."

In most cases, duplex receptacles are used for convenience outlets rated at either 15 or 20 amperes. In the kitchen, dining area, and some other locations prescribed by the NE Code, 20-ampere receptacles must be used. Those located in residential bathrooms, laundry rooms, and outside the home must also be provided with a ground-fault interrupter as an added safety precaution.

PRELIMINARY CONFERENCE

Whoever is in charge of the electrical wiring should arrange a meeting with those in charge, either the general contractor, architect, or owners, to determine the electrical conveniences desired. At this time, the designer or electrician should be prepared to make recommendations about all the electrical conveniences available. During this preliminary conference, much time can be saved if a standard form is used to collect the necessary data. Such a form should include the following:

1. *Temporary electric facilities:* Who shall furnish the temporary electric facilities during construction, and who will pay for them?

2. *Service entrance:* Overhead or underground?

3. *Service switch and panel:* Circuit breakers or switches with fuses? This same information should be determined for load centers.

4. *Branch-circuit wiring:* Does the owner or others in charge have any preference? This, of course, must be in accordance with the NE Code and local ordinances; the designer or electrician must be prepared to make recommendations.

5. *Wiring devices:* Type and color of each?

6. *Wiring-device plates:* Metal or composition? Color and finish?

7. *Low-voltage remote-control switches:* The designer must determine if the project warrants discussion of low-voltage switching of lighting circuits. If so, she or he should be prepared to discuss the advantages of this system with those in charge.

8. *Air conditioning:* What type is wanted, that is, central or through-wall room units? The mechanical designer will normally handle this phase

be used in order to plan the wiring system, feeders, and so on. He or she must also find out other necessary data about the air-conditioning system, but these other details can be found out later from the mechanical contractor.

9. *Electric heating:*   If electric baseboard heating or any other type of individual room-controlled heating is desired, the electrical designer should indicate this on the electrical drawings or sketches. However, a conflict may arise with the mechanical designer. If a forced-air duct system is planned, it will be the mechanical designer's responsibility.

10. *Signal and communications:*   The preliminary conference should be used to discuss the desired location of the telephone outlets and any other electrical conveniences chosen by the owners, such as intercom system, fire-alarm system, sound system, or central vacuum system.

After this preliminary discussion, the designer may want to verify certain conditions with the local utility company and determine the best possible route for the service drop. This is also a good time to determine any shortage of materials from suppliers that may delay the work once the project is under construction.

The best way to lay out the electrical system is to make drawings of the house. Obtain the original architectural drawings, and, with these, trace an outline of the building's floor plan on tracing paper; these drawings will eventually comprise the electrical drawings. Then sketch in all electrical outlets at their desired location on the floor plan, that is, convenience outlets, special-purpose outlets, lighting outlets, and so on. After laying out the circuitry, the entire system should be gone over to be sure that everything meets the requirements of the NE Code.

Next comes the calculation of all feeders and service-entrance equipment; later this information is transferred to the working drawing in the form of physical location on the floor plans. Wire and conduit sizes should be given in a power-riser diagram and complete information about the service equipment and load centers in the panelboard schedules.

The plumbing and mechanical drawings must be checked to determine which equipment furnished by the mechanical contractor requires electrical service and the required size of each feeder for that equipment.

After the complete sketch is finished, it should be checked thoroughly to make certain there will be no conflicts in the architectural, structural, or mechanical work and to ensure that there are no errors or omissions.

It is a rare electrical job, indeed, that does not have one or more changes in the original contract before it is completed. These electrical changes are usually the result of changes in building design, in type of equipment, or in the owner's requirements. These changes will have to be handled as the work progresses.

WIRING THE HOUSE

The wiring installation of a residence begins after the building has been framed and is under roof. Sometimes a sleeve or conduit may have to be installed in the footing or foundation so that conductors can be pulled from the outside to the inside of the building, or vice versa, but in most cases, even these are cut in after the house is under roof.

Once the house is framed, that is, all partitions are in place, but without any covering, the electricians first determine the location of the service-entrance and main panel, and then mark the location of the various outlets. Although the exact height above floor may vary slightly, convenience outlets are normally placed 15 inches above the floor to the bottom of the box, except above countertops, where they are normally positioned about 44 inches above floor level. Wall switches are normally placed 50 inches above the floor to the bottom of the box.

All outlets are plainly marked, and the appropriate-sized outlet box is secured in place at each marking, making sure that the boxes extend outward from the studs of the partitions to allow for the wall and ceiling covering, such as dry wall, plaster, or paneling.

In houses with basement space, the duplex receptacles are usually fed from underneath the floor, that is, in the ceiling space of the floor below. Holes are drilled through the floor plate under each outlet, and a wire (usually nonmetallic cable) is pulled up into each outlet box, and another pulled out of it to run to the next outlet. When the desired number of outlets have been fed, the last outlet on the series feeds directly to the panel or load center. This feeder is called a *homerun.*

Lighting outlets are normally fed from overhead in the ceiling voids, and, again, the last box on the circuit is taken "home" to the panelboard where it is connected to an overcurrent device in the panel or load center.

Switch legs and other lighting control are provided, and when all circuits have been pulled in and secured to their appropriate outlet box, connections are made in each box, usually with wire nuts, all according to the NE Code and by accepted wiring methods. The wire leads, which must be at least 6 inches long, are neatly wrapped and placed back into each outlet box to offer protection while the wall and ceiling coverings are being installed.

The locations of any special appliances, such as an electric dryer, furnace, air-conditioner, electric heating units, or water heater, are determined, and feeder circuits are run to these. In most cases, at the rough-in wiring stage, these units have not been installed, so the electrician usually rolls up the cable, to protect it during construction, and leaves it at the approximate location until the devices are in place.

The electrician will also have to determine at what stage of construction the outside wiring can be installed. This will vary, but in most cases outside wiring is installed after the final grading of the grounds.

No further wiring can be done until the wall and ceiling covering have been installed and finished, including painting. At this point, the electricians go into the home and "trim out"; but since most of the home's finishes have been completed, care must be taken not to damage any of them. At this point, wiring devices are installed and the wiring device plates set in place. All final connections are made to all special outlets, panels and load centers are made up, and all overcurrent devices are connected to the system.

During this trimming-out stage, it is also a good idea to test each circuit for any breaks or shorts that may have been caused by other trades during the period between the roughing-in and trimming stages. Sometimes a nail will be driven through an electrical cable, causing a short circuit.

When all wiring has been checked out and all circuits are complete and properly connected, the main service is energized and each circuit is turned on, one at a time, and checked out for performance.

This explanation was only a very brief sampling of the actual installation. Quite an extensive volume could be written on residential wiring alone. However, this explanation, combined with the other material presented in this book, should be a good foundation to enable one to start on residential electrical installations, especially when working with another who is already experienced in the field.

For a complete explanation of residential wiring, you might want to obtain a copy of *Residential Electrical Design* from Craftsman Book Company, 6058 Corte Del Cedro, P.O. Box 6500, Carlsbad, California 92008.

20

Planning a Commercial Wiring Installation

Electrical wiring systems for commercial buildings can range from a very small, one-room shop to huge, complex installations. For our purposes, we will cover three types of commercial buildings and categorize them as small, medium, and large.

SMALL COMMERCIAL OCCUPANCY

In planning the wiring system for a small commercial building, there are several factors to be considered before the project can be designed or before the actual installation takes place. Some of them include the following:

1. Type of general building construction.
2. Is the installation a part of a new building or a modernization of an existing one?
3. Type of ceiling, wall, and floor construction, dimensions, and the like.
4. Wiring methods.
5. Location of service, overhead or underground?
6. Location of service-entrance equipment.
7. Size of service and feeders and sizes and types of service-entrance equipment and panelboards.
8. Wiring of windows and display cases.

9. Type and installation requirements of lighting fixtures. Physical dimensions and construction of recessed lighting fixtures.

10. In the case of a modernization or complete wiring of an existing building, to what extent may the main service be used, and how much will it have to be enlarged?

Taking each of these individually, factor 1 may be determined by the working drawings and specifications, by a job site investigation, or by consulting with the owners. The same is true for factors 2 and 3.

The wiring method to use (4) may be dictated by the working drawings or specifications. If not, the wiring method to use should comply with the latest edition of the NE Code and/or local ordinances.

The location of service equipment (5 and 6) may be indicated on the working drawings or the local power company may have to be consulted to determine the best location. Locating the service equipment is often left to the electrician or contractor to decide, but if working drawings have been made, they should be followed.

Sizing the electric service (7) requires calculations as discussed in Chapter 13 or the designer may have indicated the service size on the working drawings.

Factor 8 can be answered by either the working drawings or by consulting the architect or owner, or it can be worked out between the trades on the job.

Lighting fixtures (9) should be preselected by either the designer, owner, or architect. Installation details may be found in manufacturers' catalogs and shop drawings.

Factor 10 can be determined either from the working drawings or by a job site investigation.

In general, the designer or electrician performing the work will calculate the total load for the building, determine the number of branch circuits required and service-entrance size, along with feeders, service-entrance equipment, and panelboards. The number of outlets will be determined along with their location. Illumination levels are calculated and then lighting fixtures are selected to provide the required illumination.

Continue by noting connections for any special equipment, such as water heater or air conditioning. Also determine the requirements for security/fire-alarm system, display case connections, and the like.

Determine the lengths of all branch circuit, service, and feeder runs and list the wire size for each. Account for service-entrance equipment and any other major pieces of equipment requiring electrical connections.

The preceding information should provide a good summary of the material needed for the job to be used in estimating costs and the number of men required for the installation and to aid in ordering the required material.

In most cases, small commercial projects utilize rigid steel conduit for the service-entrance regardless of whether it is overhead or underground. Check with the local power company to find out exactly what is required of the contractor or electricians doing the work. Either rigid or EMT conduit is used for all wiring below grade and embedded in concrete slabs. Either EMT or type AC cable is normally used for wiring above grade.

Be extremely careful of any wiring that may be installed in hazardous locations, such as in commercial service stations around the gas pumps and in the garage area. See Chapter 24 for descriptions of other hazardous locations.

One main point of concern with this type of project, as well as with most other electrical installations, is to plan the job well so as to perform the work in the shortest possible time, yet keep the quality high and in a workmanlike manner. Other trades should not be held up in performing their work, and the electrical worker must plan and work accordingly. For example, before the concrete floor may be poured and finished, all conduit, boxes, and fittings must be installed by the electricians. When the ground is graded, wire mesh installed, and the like, the electrical workers usually have only a certain amount of time to complete their portion of the work. Make certain all necessary material is at hand on the job site well before the installation will take place. Have the working drawings or at least a sketch at hand to go by, and work efficiently when the time comes. Also double-check each homerun, circuit, and outlet box location, as once the cement is poured, it would be quite costly to make any changes under it.

At least one electrician should be present during the pouring to ensure that none of the electrical system is damaged; if it is, the damage should be corrected immediately before the concrete sets up.

MEDIUM COMMERCIAL BUILDING

A medium-sized commercial occupancy is planned much like the smaller building just described, except there will be more circuits, a larger service entrance, and so on. In nearly all cases, working drawings will be provided by an architectural-engineering firm to consult during the job. If engineer's drawings are not provided, the electrical contractor should provide some type of layout to be followed. Not only do such drawings aid the workers as the job progresses, but they also give a means of knowing what has been installed at a later date while the building's electrical system is being maintained or repaired.

Depending on the use of the building, the NE Code or local ordinances may require a different wiring method than would be required in a smaller building.

LARGE COMMERCIAL BUILDING

Most larger commercial buildings utilize a 480/277-volt Y-connected service-entrance; all heavy equipment, such as compressors for air conditioning, are designed for use on 480 volts; electric discharge lighting is all designed for operation on 277-volt, single-phase circuits; and dry transformers are required to obtain 120 volts for convenience outlets; and other outlets use 240 and/or 120 volts.

Factors affecting wiring systems in large commercial buildings include the following:

1. Type of building construction, that is, masonry, reinforced concrete, wood frame, and the like.
2. Type of floor, ceiling, and partition construction, height of ceiling, space above ceiling, space under floor, and the like.
3. Wiring methods, type of raceway, sizes of conductors.
4. Type of service-entrance equipment.
5. Type of service and location of service conductors.
6. Connections for equipment not furnished by the electrical contractor but requiring electric service.
7. Type and construction of lighting fixtures, hangers, and supports affecting assembly and installation. Types of lamps.
8. Type and dimensions of floodlighting supporting poles, floodlights and mounting brackets, and so on.
9. Ground conditions affecting trenching for parking lot lighting.

The majority of the factors can be determined by examining the working drawings and specifications, as any commercial building of this size will have a detailed, engineered set of drawings and specifications. If not, the contractor will have to have the system designed and working drawings made to aid the workers on the job. Building inspectors often also require that they be supplied with at least one set of drawings and specifications for use in their office and to check against the actual installation.

In many cases, it is also a good idea for the contractor to examine the job site conditions prior to bidding or beginning the electrical installation.

A complete take-off of materials will be required for this size of installation for the purpose of estimating the cost of construction, as well as for ordering material and scheduling it for use at the job site.

In many instances, consulting engineers will prepare drawings that leave out much detail, requiring the contractor or his personnel to do extensive research to determine exactly what is taking place. For example, a main distribution panelboard may be indicated on the drawings only by a symbol on the floor plan layout and a catalog number of the equipment. A better drawing, however, will have a complete power riser diagram to supplement the floor plan drawing, showing conduit sizes, wire sizes, number of conductors, and so on. The person installing the system, when only meager symbols are used, usually will have to make a rough layout of the installation and list all details before materials can be ordered or the installation started. Calculations will have to be performed to determine wire size, limit voltage drop, size of conduit, and the like. All these details should be worked out prior to starting the electrical installation.

While commercial electrical installations may vary considerably in detail, in general the majority of them will follow a definite pattern. For example, each will have a service entrance, a distribution panelboard, lighting, and convenience outlets. Furthermore, nearly all will have emergency lighting and signal systems. All will have branch circuits, feeders, and the like.

Therefore, when the electrical technician is called on to design or install a commercial electrical installation, there should really be no "strange" jobs once he or she has worked on a few commercial installations. Then, by following sound basic planning techniques and giving careful attention to details, the trained technician should never be completely stumped, even on jobs of a type that have not been previously handled.

A certain amount of research will have to be done on all electrical jobs. Even seasoned professionals constantly refer to reference material for practically every new project. For example, while the professional engineer may remember the required footcandle level of, say, an office area, he or she will have to refer to manufacturers' catalogs to obtain the illuminating characteristics of certain lighting fixtures. Tables will be consulted to determine voltage drop on various sizes of wire over a given distance and carrying a certain load. Short-circuit calculations will be made to specify the required overcurrent protection—just to name a few. However, the pattern or sequence in which these unknowns are determined is practically the same on every commercial job.

Workers on the job have a further responsibility. While the better engineered drawings and specifications are coordinated to a certain extent with the architectural drawings and the work of other trades, none can be absolutely complete all the time. It is up to the workers on the job to be sure that conduit runs will not interfere with the equipment of other trades. Furthermore, they must make certain that the electrical equipment will not weaken the structural members of the building. The electricians are also required to lay out the circuit runs to use the least amount of material, yet see that the finished job is done in a professional manner.

Sometimes it becomes necessary to vary from the working drawings considerably during the installation, but before doing so, the consulting engineer or architect should be consulted for approval.

In summary, regardless of the technician's position—designer, electrician, foreman, or whatever—a certain amount of planning is required on all commercial electrical installations. This planning is begun, for the person's respective responsibility, before any work is started, and then continues on a day-to-day basis until the project is completed. Then a certain amount of planning is required to perform the final tests of the entire system.

Even on projects with detailed engineered drawings and specifications, planning and coordination during the construction phase are still necessary. Certain phases of the electrical installation will have to be carefully planned so as not to hold up any other trades from doing their work. Local inspectors will have to be notified at certain times so that they can inspect the work before it is covered up. Material and tools will have to be ordered so that they will be on the job site when needed. The design group will have to make periodic checks to ensure that the equipment specified is in fact being used, while the workers or the foreman on the job will have to make certain that installed equipment is not damaged by other trades while performing their respective work.

While other factors beyond your control may adversely affect the final electrical installation, job planning and carrying out this plan are largely the basis on which the work will be performed. Be certain that the planning is done on a sound basis.

21

Planning an Industrial Wiring Installation

The same basic principles of job planning covered in Chapters 19 and 20 also apply to a certain extent to industrial applications. Larger currents, requiring larger wire and conduit, will be encountered; higher voltages will normally be used; three-phase in addition to single-phase will invariably be used; and different types of materials and equipment may be involved.

FACTORS TO CONSIDER

Factors affecting the planning of an electrical installation for an industrial plant include the following:

1. New structure or modernization of an existing one.
2. Type of general building construction; that is, masonry, reinforced concrete, steel frame.
3. Type of floor, ceiling, partition, roof, and so on.
4. Type and voltage of service entrance, transformer connections, and underground or overhead.
5. Type and voltage of distribution system for power and lighting.
6. Type of required service equipment, such as unit substation, transformer bank, and the like.
7. Type of distribution system, including step-down transformers.
8. Who is responsible for furnishing the service-entrance and distribution equipment, the power company or plant?
9. Wiring methods, types of raceways, special raceways, busways, and the like.

10. Types of power-control equipment and extent of the worker's responsibility for connection thereto.
11. Furnishing of motor starters, controls, and disconnects.
12. Extent of wiring to be installed on machine tools.
13. Extent of wiring and connections to electric cranes and similar apparatus.
14. Type and construction of lighting fixtures, hangers and supports, and the like.
15. Extent of floodlighting. Type and dimensions of floodlighting supporting poles and mounting brackets.
16. Extent of signal and communication systems.
17. Ground conditions affecting the installation of underground wiring.
18. The size, type, and condition of existing wiring systems and services in case of a modernization project.
19. Whether the plant will be in use during the electrical installation.
20. Allowable working hours when an occupied building is being rewired.
21. Labor disputes.

Items 1 through 3 can be determined by studying the architectural drawings of the facility or, in the case of a modernization project, by a job site investigation.

Item 4 is usually worked out between the engineers and the local utility company, while items 5 through 10 are determined from the electrical drawings and specifications.

The electrical drawings as well as the mechanical drawings should be consulted to determine the condition of items 10 and 11. Also consult the special equipment sections of the written specifications. In most cases, it is the responsibility of the trade furnishing motor-driven equipment to also furnish the starters and controls. The electrical contractor then provides an adequate circuit to each.

Item 12 can be determined by consulting the special equipment section of the written specifications and also by referring to the shop drawings supplied by the trade furnishing the equipment. It may be necessary to contact the manufacturer of the equipment in some cases.

On most projects, crane installation is a specialized category and installed by specialists in this field. All control work for the crane is normally done by the contractor who furnishes the crane. However, electrical workers will be required to furnish a feeder circuit from the main distribution panel to supply power for motors and controls to operate the crane. The extent of this work should be carefully coordinated between the trades involved.

Items 14 through 16 can be determined by consulting the electrical drawings and specifications for the project.

While items 17 through 19 might be covered to a certain extent in the project specifications, in most cases this information is obtained by a job site investigation.

Item 20 should be called out in the general specifications, but it may be necessary to hold a conference with the owners to determine exact conditions. Item 21 is determined by referring to the local labor agreement, by contacting the local union, or from past experience.

PLANNING AND COORDINATION

Even with carefully engineered drawings, the person in charge of the electrical installation in an industrial occupancy must still do much planning and coordination to carry out the work in the allotted amount of time. One problem that has existed in the past has been a variation of interpretation of the code requirements by two or more inspection authorities having coinciding jurisdiction over the same job. Therefore, at the planning stage, the superintendent or foreman should meet with all inspection authorities having jurisdiction to settle any problems at the outset.

Most industrial plants will have one or more wiring installations in hazardous locations. Therefore, those in charge should frequently consult the NE Code to ascertain that all wiring is installed in a safe manner. Provisions must also be made to isolate the hazardous areas from those not considered hazardous.

Drawings from consulting engineers will vary in quality, and in most cases the wiring layout for a hazardous area is little different than the layout for a nonhazardous area. Usually the only distinction is a note on the drawing or in the specifications stating that the wiring in a given area or room shall conform to the NE Code requirements for hazardous locations. Rarely do the working drawings contain much detail of the system, leaving much of the design to the workers on the job. Therefore, the electrical foreman must study these areas very carefully, at the same time consulting the NE Code and other references, to determine exactly what is required.

Sometimes the electrical contractor will have draftsmen prepare special drawings for use by his personnel in installing systems in such areas. If time permits, this is probably the best approach as it will save much time in the field once the project has begun. In either case, preparing drawings or determining the requirements at the job site, considerable care must be made during this planning stage. One reason is because considerable damage can be done to life and property if the system installed is faulty, and a second is because explosion-proof boxes, fittings, and equipment are very expensive and vary in cost for different types, sizes, and hub entrances, and relatively considerably more labor is required than for nonhazardous installations.

For other than very simple systems, it is advisable to make detailed wiring layouts of all wiring systems in hazardous locations, even if it is only in the form of sketches and rough notes.

The typical plant wiring system will entail the connection of a power supply to many motors of different sizes and types. Sometimes the electrical contractor will be responsible for furnishing these motors, but in most cases they will be handled on special order from the motor manufacturers and will be purchased direct by the owners or by special equipment suppliers. Some factors involved in planning the wiring for electric motors include the following:

1. The type, size, and voltage of the motor and related equipment.
2. Who furnishes the motor, starter, control stations, and disconnecting means?
3. Is the motor separately mounted or an integral part of a piece of machinery or equipment?
4. Type and size of junction box or connection chamber on the motor.
5. The extent of control wiring required.
6. The type of wiring method of the wiring system to which the motor is to be connected, that is, conduit and wire, bus duct, trolley duct. Is the motor located in a hazardous area? If so, what provisions have been made to ensure that it will be wired according to the NE Code?
7. Who mounts the motor?
8. The physical shape and weight of the motor.

Obtaining all the above information will facilitate the installation of all electric motors on the job by ensuring that proper materials will be available for the wiring and that no conflicts will arise between trades.

The installation of transformers and transformer vaults is another type of work that is frequently encountered in industrial wiring. A transformer vault, for example, is representative of the type of installation situation when, within a small area of the building and comprising a specialized section of the wiring system, a relatively small quantity each of a number of different items of equipment and materials is required. In many instances, even when working drawings and specifications are provided by consulting engineering firms, the vault will not be completely laid out to the extent that workers can perform the installation without further planning or questions. The major transformers, disconnects, and similar devices along with a one-line schematic diagram may be all that the drawings show. In such instances the foreman or workers must make a rough layout of the primary and secondary, indicating the necessary supports, supporting structures, connections, control and metering wiring, and the like. This calls for experienced knowledge and the ability to visualize the complete installation on the part of the person during the layout work and supervision.

CABLE TRAY SYSTEMS

Cable tray systems are frequently used in industrial applications, and all electrical technicians involved in such work should be thoroughly familiar with the design and installation of such systems.

In general, a cable tray system must afford protection to life and property against faults caused by electrical disturbances, lightning, failures that are a part of the system, and failure of equipment that is connected to the system. For this reason, all metal enclosures of the system, as well as noncurrent-carrying or neutral conductors, should be bonded together and reduced to a common earth potential.

There is a frequent tendency to become lax in supplying the installation supervisor with definite layouts for a cable tray system. Under these conditions, considerable time is consumed in arriving at final decisions and definite routings before the work can proceed. On the other hand, it is often possible for one person to predetermine these layouts and save many hours of field erection time, provided careful planning is carried out.

For economical erection and satisfactory installation, working out the details of supports and hangers for the system is the job of the system designer and should not be left to the judgment of a field force not acquainted with the loads and forces to be encountered. Also, all types of supports and hangers should permit vertical adjustment, along with horizontal adjustment where possible. This can be accomplished by the use of channel framing, beam clamps, and threaded hanger rods.

All cable tray systems of appreciable size require a considerable quantity of hanger clips, support channels, hangers, and the like, in addition to cable tray items, which are intended for specific locations and may vary considerably in specifications. Too much emphasis, therefore, cannot be given to the necessity of packing, delivering, and receiving the material under definite and clear records so that the various items will be readily available when required, and also will not be confused with materials that are similar in appearance. All these factors should be considered during the planning stage.

When installing the cables, proper precaution must be taken to avoid damaging the cables. A complete line of installation tools is available, developed through field experience, for pulling long lengths of cable up to 1000 feet or longer. These tools save considerable installation time.

Short lengths of cable can be laid in place without tools or pulled with a basket grip. Long lengths of small cable, 2 inches or less in diameter, can also be pulled with a basket grip. Larger cables, however, should be pulled by the conductor and the braid, sheath, or armor. This is done with a pulling eye applied at the cable factory or by tying the conductor to the eye of a basket grip and taping the tail end of the grip to the outside of the cable.

In general, the pull exerted on the cables pulled with a basket grip, not attached to the conductor, should not exceed 1000 pounds. For heavier pulls, care should be taken not to stretch the insulation, jacket, or armor beyond the end of the conductor nor bend the ladder, trough, or channel out of shape.

The bending radius of the cable should not be less than the values recommended by the cable manufacturer, which range from four times the diameter for a rubber-insulated cable 1-inch maximum outside diameter without lead, shield, or armor, to eight times the diameter for interlocked armor cable. Cables of special construction such as wire armor and high-voltage cables require a larger radius bend.

Best results are obtained in installing long lengths of cable up to 1000 feet with as many as a dozen bends by pulling the cable in one continuous operation at a speed of 20 to 25 feet per minute. It may be necessary to brake the reel to reduce sagging of the cables between EZ rolls, which are devices designed especially for pulling cables in cable trays.

The pulling line diameter and length will of course depend on the pull to be made and construction equipment available. Winch and power unit shall be of adequate size for the job and capable of developing the high pulling speed required for best and most economical results.

22

Wiring for Mobile Homes and Recreation Vehicle Parks

During the past decade, mobile home and trailer parks have increased in number to the point where the NFPA found it necessary to include detailed wiring instructions covering such operations. Article 550 and Article 551 of the NE Code cover the electrical conductors and the equipment installed within or on mobile homes and recreation vehicles, as well as the means of connecting the units to an electrical supply.

SIZING ELECTRICAL SERVICES FOR MOBILE HOMES

The electrical service for a mobile home may be installed either underground or overhead, but the point of attachment must be a pole or power pedestal located adjacent to but not mounted on or in the mobile home. The power supply to the mobile home itself is provided by a feeder assembly consisting of not more than three mobile home power cords, each rated for at least 50 amperes or a permanently installed circuit.

The NE Code gives specific instructions for determining the size of the supply-cord and distribution-panel load for each feeder assembly for each mobile home. The calculations are based on the size of the mobile home, the small-appliance circuits, and other electrical equipment that will be connected to the service.

Lighting loads are computed on the basis of the mobile home's area: width times length (outside dimensions exclusive of coupler) times 3 watts per square foot.

$$\text{Length} \times \text{width} \times 3 = \underline{\hspace{2cm}} \text{ lighting watts}$$

Small-appliance loads are computed on the basis of the number of circuits times 1500 watts for each 20-ampere appliance receptacle circuit.

Number of circuits $\times$ 1500 = _____ small-appliance watts

The sum of the two loads gives the total load in watts. However, there is a diversity (demand) factor that may be applied to this total in sizing the service and power cord. The first 3000 watts (obtained from the previous calculation) is rated at 100 percent. The remaining watts should be multiplied by a demand factor of 0.35 (35 percent). The total wattage so obtained is divided by the feeder voltage to obtain the service size in amperes.

If other electrical loads are to be used in the mobile home, the nameplate rating of each must be determined and entered in the summation. Therefore, to determine the total load for a mobile home power supply, perform the following calculations:

1. Lighting and small appliance load, as discussed previously.
2. Nameplate amperes for motors and heater loads, including exhaust fans, air conditioners, and electric heaters. Since air conditioners and heaters will not operate simultaneously, only the larger of the two needs to be included in the total load figures. Multiply the largest motor nameplate rating by 1.25 and add the answer in the calculations.
3. Total of nameplate amperes for any garbage disposals, dishwashers, electric water heaters, clothes dryers, cooking units, and the like. Where there are more than three of these appliances, use 75 percent of the total load.
4. The amperes for free-standing ranges (as distinguished from separate ovens and cooking units) by dividing the values shown in Table 22-1 by the voltage between phases.
5. The anticipated load if outlets or circuits are provided for other than factory-installed appliances.

TABLE 22-1.   Power Demand Factors for
Free-Standing Electric Ranges

Nameplate Rating (watts)	Use
10,000 or less	80% of rating
10,001-12,500	8,000 watts
12,501-13,500	8,400 watts
13,501-14,500	8,800 watts
14,501-15,500	9,200 watts
15,501-16,500	9,600 watts
16,501-17,500	10,000 watts

To illustrate this procedure for determining the size of the electrical service and power cord for a mobile home, assume that a mobile home is 70 feet by 10 feet; has three portable appliance circuits; a 1200-watt air conditioner; a 200-watt, 120-volt exhaust fan; a 1500-watt water heater; and a 8000-watt electric range. The load is calculated as shown in Table 22-2.

Based on the higher current for either phase, a 60-ampere power cord should be used to furnish electric power for the mobile home. The service should be rated for a minimum of 60 amperes and be fused accordingly.

TABLE 22-2.

Lighting and Small-Appliance Load	Watts
Lighting load: 70 × 10 × 3 W/ft²	2100
Small-appliance load: 1500 W/circuit × 3 circuits	4500
	6600
First 3000 at 100 percent	3000
Remainder: 6600 − 3000 = 3600 at 35%	1260
	4260

$$\frac{4260 \text{ watts}}{240 \text{ volts}} = 17.75 \text{ amperes per phase}$$

Large-Appliance Load	
1500 W (water heater) ÷ 240 V	6.25 A/line
200 W (fan) ÷ 120 V (one phase)	1.66 A/phase
1200 W (air conditioner) ÷ 240 V	5.00 A/line
8000 W (range) × 0.8 (diversity factor) ÷ 240 V	26.66 A/line

	Amperes Per Phase	
Summary	A	B
Lighting and appliance	17.75	17.75
Water heater	6.25	6.25
Fan	1.66	——
Air conditioner	5.00	5.00
Range	26.66	26.66
Total	57.32	55.66

TYPES OF EQUIPMENT

Weatherproof electrical equipment for mobile homes, mobile home parks, and similar outdoor applications are available from many sources. One of the major suppliers of weatherproof power outlets is Midwest Electric Products, Inc., Mankato, Minnesota. The reader should obtain one of their free catalogs and study the many types of mobile home utility power outlets and service equipment manufactured by this company.

Receptacle configurations used in mobile home and recreation vehicle applications are shown in Fig. 22-1. Receptacles 1, 2, 4, and 8 are the most commonly used in mobile home and recreation vehicle parks.

Power units for use in mobile homes and recreation vehicles vary in design with fuse or circuit-breaker projection, attached meter sockets, mounting and junction posts, and special corrosion-resistant finish for ocean-side areas. The latter are used for boats where the outlets are located along the docks and power is leased by the dock owners on a daily basis. In addition to the standard units, manufacturers build equipment to meet special requirements. Merely write the manufacturer, and they will supply you with a price and delivery date on the equipment.

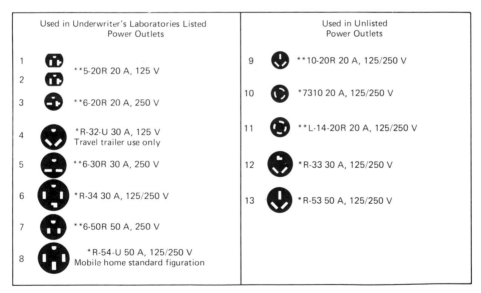

Used in Underwriter's Laboratories Listed Power Outlets		Used in Unlisted Power Outlets	
1	**5-20R 20 A, 125 V	9	**10-20R 20 A, 125/250 V
2			
3	**6-20R 20 A, 250 V	10	*7310 20 A, 125/250 V
4	*R-32-U 30 A, 125 V Travel trailer use only	11	**L-14-20R 20 A, 125/250 V
5	**6-30R 30 A, 250 V	12	*R-33 30 A, 125/250 V
6	*R-34 30 A, 125/250 V	13	*R-53 50 A, 125/250 V
7	**6-50R 50 A, 250 V		
8	*R-54-U 50 A, 125/250 V Mobile home standard figuration		

*Midwest receptacle number
**NEMA receptacle configuration number

Figure 22-1   Receptacle configurations.

Where more than one mobile home is to be fed, a power outlet and service equipment mounting cubicle (section shown in Fig. 22-2) is ideal. The bus bars in this unit accommodate wire sizes to 600 MCM for a 400-ampere capacity. Therefore, either two 200-ampere units or four 100-ampere units may be mounted on the cubicle for feeding mobile home units underground. The main service should also enter from underground.

Standard 15- or 20-ampere, 120-volt receptacles may be converted for use with standard 30-ampere, 120-volt travel trailer caps by use of the adapter shown in Fig. 22-3.

The power outlet mounting post shown in Fig. 22-4 is very popular for travel trailer parks and marinas. A cast-aluminum mounting base is provided to mount the power outlet on. The installer provides a length of 2-inch rigid conduit of the desired length. This is for underground installations. Note the conductors feeding in and out of the bottom of the conduit.

A side view of another mounting post and wire trough for travel trailer parks where the installation requires the power outlet to be mounted remote from the meter socket is shown in Fig. 22-5. A mounting post for mobile home applications is shown in Fig. 22-6. This post accommodates one or two combination meter-power units back-to-back on the posts. When the power units are ordered from the factory prewired, the entire unit is ready to be secured in the ground and only the connections to the terminal bar have to be made.

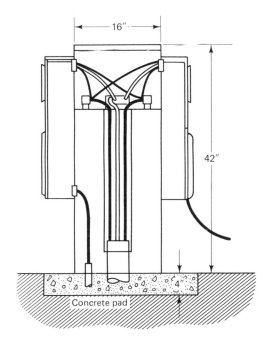

Figure 22-2   Mounting cubicle for accommodating up to four mobile homes.

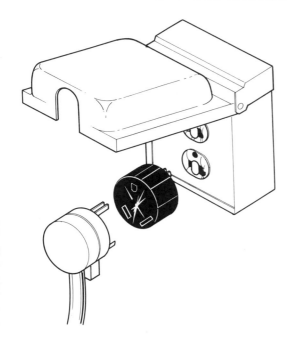

Figure 22-3   Travel trailer adapters are available to adapt the standard, 30-ampere, 125-volt travel trailer cap to any standard, 15- or 20-ampere, 125-volt outlet.

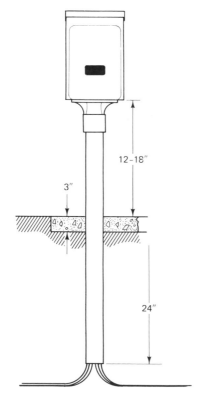

Figure 22-4   A popular power-outlet mounting post.

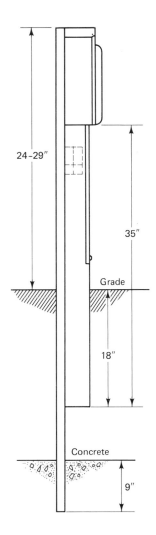

Figure 22-5   Side view of another type of mounting post used in parks for travel trailers.

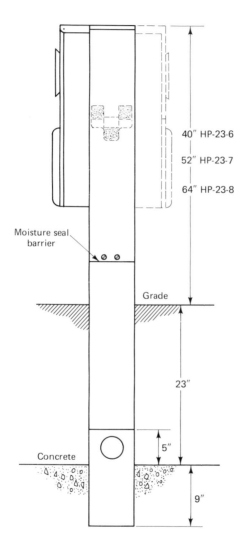

40" HP-23-6

52" HP-23-7

64" HP-23-8

Moisture seal
barrier

Grade

23"

5"

Concrete

9"

Figure 22-6   Mounting post for
mobile home applications.

INSTALLATION METHODS

There are certain requirements in the NE Code as well as accepted methods for installing mobile home services and feeders. The electrician should become familiar with these prior to beginning an installation. Some of the more commonly used methods are discussed next.

A typical mobile home park service pole and meter installation is shown in Fig. 22-7. This particular arrangement is designed to serve four mobile home units.

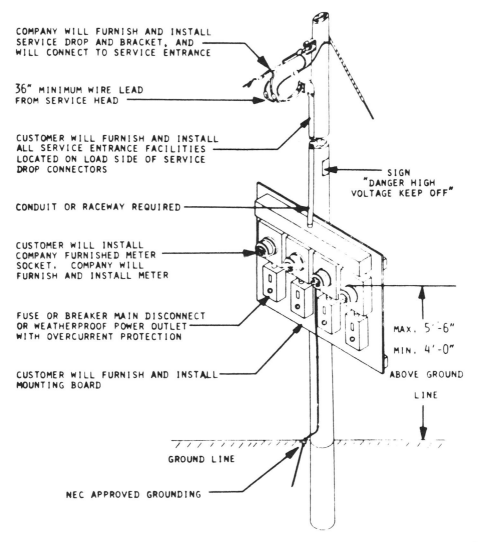

COMPANY WILL FURNISH AND INSTALL SERVICE DROP AND BRACKET, AND WILL CONNECT TO SERVICE ENTRANCE

36" MINIMUM WIRE LEAD FROM SERVICE HEAD

CUSTOMER WILL FURNISH AND INSTALL ALL SERVICE ENTRANCE FACILITIES LOCATED ON LOAD SIDE OF SERVICE DROP CONNECTORS

SIGN "DANGER HIGH VOLTAGE KEEP OFF"

CONDUIT OR RACEWAY REQUIRED

CUSTOMER WILL INSTALL COMPANY FURNISHED METER SOCKET. COMPANY WILL FURNISH AND INSTALL METER

FUSE OR BREAKER MAIN DISCONNECT OR WEATHERPROOF POWER OUTLET WITH OVERCURRENT PROTECTION

MAX. 5'-6"

MIN. 4'-0"

CUSTOMER WILL FURNISH AND INSTALL MOUNTING BOARD

ABOVE GROUND

LINE

GROUND LINE

NEC APPROVED GROUNDING

Figure 22-7   Typical mobile home park service. (Courtesy Potomac Edison Power Co.)

Before beginning an installation, the customer or the electrician performing the work should consult with the local power company for the method of serving the mobile home park and for the location of service-entrance poles. Power company regulations vary from area to area, but the power company will furnish and set the pole in most cases. However, the electrician must obtain permission from the power company before performing any work on facilities on the poles.

Once a definite plan has been settled on, the electrician should obtain a piece of 1/2-inch-thick plywood of sufficient size to hold the service equipment (i.e., a wire trough, meter bases, and weatherproof power outlets as shown in Fig. 22-7). This piece of plywood should be primed with paint, and a final coat of wood preservant then applied. Two pieces of 2- by 4-inch timbers are spiked or otherwise secured to the sheet of plywood for reinforcement before the entire assembly is spiked to the pole. The wood backing should be arranged so that the meters will be no more than 5 feet 6 inches above the ground nor less than 4 feet when they are installed.

Up to six power outlets may be fed from one service without need for a disconnect switch to shut down the entire service. However, if more than four power outlets are assembled on one piece of plywood, the arrangement shown in Fig. 22-7 will not provide adequate support. Another short pole should be installed at a distance from the service pole so that the sheet of plywood can be secured to both poles (Fig. 22-8) for added support.

With the plywood backing secured in place, a wire trough, sized according to Article 374-5 of the NE Code, should be installed at the very top of the board as shown in Fig. 22-7. The wire trough (auxiliary gutter) should not contain more than 30 current-carrying conductors nor should the sum of the cross-sectional areas of all contained conductors at any cross section exceed 20 percent of the interior cross-sectional area of the gutter. The auxiliary gutter should be approved for outdoor use.

The meter bases may usually be obtained from the power company but must be installed by the electrician. Once the entire installation is complete and has been inspected, the power company will install the meters. The connections of the meter bases to the wire trough are made with short, rigid conduit nipples using locknuts and bushings. Although straight nipples are often used for these connections, an offset nipple (Fig. 22-9) usually does a better job.

Weatherproof fuse or circuit-breaker disconnects are installed directly under the meter bases, again by means of conduit nipples. A weatherproof, 50-ampere mobile home power outlet with overcurrent protection is also used quite often.

The service mast comes next and should consist either of rigid metallic conduit or of EMT with weatherproof fittings. Once installed and secured to the pole with pipe straps, the service-entrance conductors may be pulled into the conduit and out into the wire trough. An approved weatherhead is then

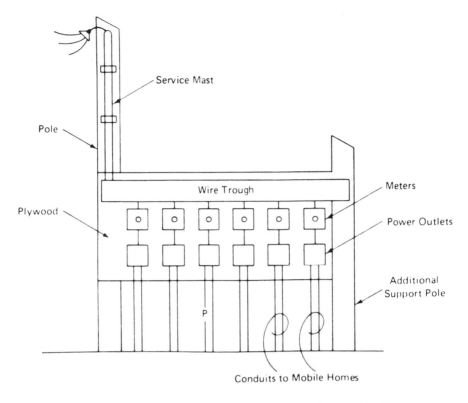

Figure 22-8   Method of installing an additional short pole adjacent to the service pole for added support.

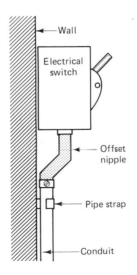

Figure 22-9 Offset nipple used to connect pieces of electrical equipment.

installed on top of the mast, and at least three feet of service conductors should be left for the power company to make their connections.

With the service-entrance conductors in place, meter taps are made to the service conductors in the trough. All such splices and taps made and insulated by approved methods may be located within the gutter when the taps are accessible by means of removable covers or doors on the wire trough. The conductors, including splices and taps, must not fill more than 75 percent of the gutter area (Article 374-8a). These taps must leave the gutter opposite their terminal connections, and conductors must not be brought in contact with uninsulated current-carrying parts of opposite polarity.

The taps in the auxiliary gutter go directly to the line side of the meter bases. Once secured, the load side of the meter bases is connected to the disconnects or power outlets. All wiring should be sized according to the NE Code.

Most water supplies for mobile homes consist of PVC (plastic) water pipe and, therefore, cannot provide an adequate ground for the service equipment. In cases like these, a grounding electrode, such as a 3/4-inch by 8-foot ground rod driven in the ground near the service equipment is used. A piece of bare copper ground wire is connected to the ground rod on one end with an approved ground clamp, and the other end is connected to the neutral wire in the auxiliary gutter. This wire must be sized according to Table 250-94a of the NE Code.

When all the work is complete, the service installation should be inspected by the local electrical inspector. The power company should then be notified to provide final connection of their lines.

SIZING ELECTRICAL SERVICES AND FEEDERS FOR PARKS

A minimum of 75 percent of all recreation vehicle park lots with electrical service equipment must be equipped with both a 20-ampere, 125-volt receptacle and a 30-ampere, 125-volt receptacle. The remainder with electrical service equipment may be equipped with only a 20-ampere, 125-volt receptacle.

Since most travel trailers and recreation vehicles built recently are equipped with 30-ampere receptacles, an acceptable arrangement is to install a power pedestal in the corner of four lots so that four different vehicles can utilize the same pedestal. Such an arrangement requires three 30-ampere receptacles and one 20-ampere receptacle to comply with Section 551-44 of the NE Code. A wiring diagram showing the distribution system of a park electrical system serving 20 recreation vehicle lots is shown in Fig. 22-10.

Electric service and feeders must be calculated on the basis of not less than 3600 watts per lot equipped with both 20- and 30-ampere supply facilities and 2400 watts per lot equipped with only 20-ampere supply facilities. The demand factors set forth in Table 551-44 of the NE Code are the minimum allowable factors that may be used in calculating load for service and feeders.

Example 1:

Park area A has a capacity of 20 lots served by electricity; park B has 44. Find:
1. The diversity (demand) factor of area A
2. The diversity (demand) factor of area B
3. The total demand of area A
4. The total demand of area B

Solution:
1. The diversity factor is 26 percent, read directly from Table 551-44 of the NE Code.
2. The diversity factor is 24 percent, read directly from Table 551-44 of the NE Code.
3. Since each lot is calculated on the basis of 3600 watts, the total demand is $20 \times 3600 \times 0.26 = 18{,}720$ watts.
4. The total demand is $3600 \times 44 \times 0.24 = 38{,}016$ watts (total demand).

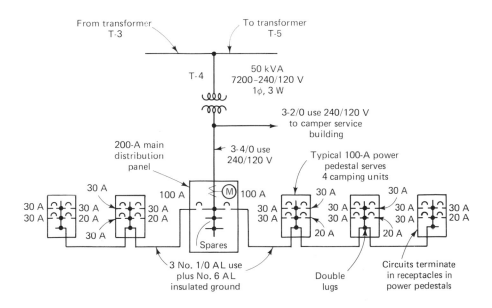

Figure 22-10   Wiring diagram showing the distribution system of a trailer park.

23

Farm Wiring

Electricity on the farm has progressed in less than 50 years from 32-volt dc battery systems that powered mostly lighting, to ultramodern 120/240-volt ac systems used to operate numerous time-saving devices around the farm. Quartz and HID lighting fixtures have now replaced the inefficient incandescent lamps that once were sparingly placed in essential locations, allowing much work to be done after daylight hours. In fact, farm wiring has become so specialized that some electrical contractors work only on farm electrical installations, and have their hands full at that!

Adequate wiring, properly planned and installed, provides safe lighting for every task, reduces eyestrain and fatigue, and reduces the time required for household and farm work and chores. Electricity provides convenient power for milking machines, milk coolers, feed grinders, silo loaders, chicken brooders, and grain elevators, just to name a few. But these conveniences and comforts are not provided without adequate planning and wiring.

FARM SERVICE ENTRANCES

The service for the farm begins at the secondary terminals of the transformers connected to the utility company's lines. Two types are usually readily available: single phase and three phase. Of these two, single phase is by far the most common type found on farms throughout the United States and Canada. It will provide satisfactory power for lighting circuits and convenience outlets and for single-phase motors rated up to about 7-1/2 horsepower. When motors larger than 7-1/2 horsepower are to be used, a three-phase delta service is almost always provided.

The single-phase, three-wire service is provided by a transformer as shown in the wiring diagram in Fig. 23-1. Note that two single-phase primary conductors are connected to the primary winding of the transformer through fuse cutouts; lightning arrestors are also provided. The secondary windings have three conductors connected to the transformer terminals, two "hot" wires and a neutral, the latter of which is always grounded. The voltage between the two hot wires is 240 volts, while the voltage between the neutral and any one hot wire is 120 volts.

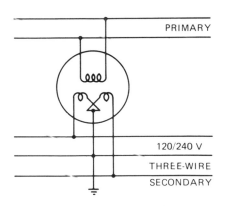

PRIMARY

120/240 V

THREE-WIRE

SECONDARY

Figure 23-1   Diagram of a single-phase, three-wire, 120/240-volt transformer connection, typical of those used for most farm service entrances. (Courtesy Potomac Edison Power Co.)

When possible, the transformer should be located near the center of the electrical load to reduce the size and length of feeder wires and thus reduce the cost of the installation. A typical farm electric service is shown in Fig. 23-2.

Three-phase service is quite common when electric motors are used that are in excess of 7-1/2 horsepower. Motors driving large irrigation pumps, feed grinders, crop-drying fans, and the like often require 10 horsepower or more. Voltage regulation with three-phase service is better, and the cost of three-phase motors is generally less than for single-phase motors of the same horsepower rating.

Two types of three-phase service are available in most areas: four-wire wye and four-wire delta. Of the two, the four-wire delta seems to be used the most, mainly because 240-volt single-phase motors and appliances may also be operated satisfactorily on this service, whereas the wye type supplies only 208 volts between phases. However, caution must be employed with the four-wire delta system to make sure the "high leg," which supplies about 180 volts, is not mistakenly connected to a 120-volt outlet. See Figs. 23-3 and 23-4 for connections of these systems.

Also, 208-volt appliances and motors are available for use on the wye system, and where a number of 120-volt motors and appliances are in use, the wye is easier to balance than the delta system. There is also less chance of an improper connection that might be made by an inexperienced person.

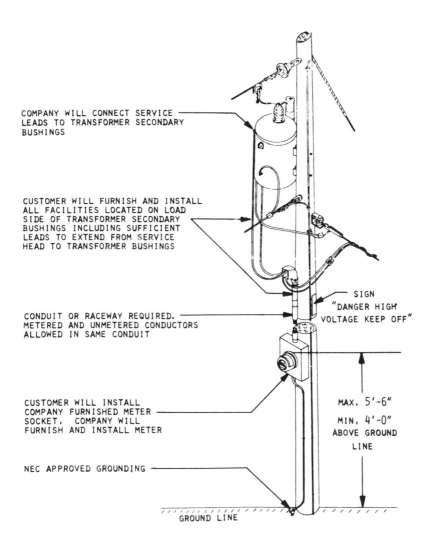

COMPANY WILL CONNECT SERVICE
LEADS TO TRANSFORMER SECONDARY
BUSHINGS

CUSTOMER WILL FURNISH AND INSTALL
ALL FACILITIES LOCATED ON LOAD
SIDE OF TRANSFORMER SECONDARY
BUSHINGS INCLUDING SUFFICIENT
LEADS TO EXTEND FROM SERVICE
HEAD TO TRANSFORMER BUSHINGS

SIGN
"DANGER HIGH
VOLTAGE KEEP OFF"

CONDUIT OR RACEWAY REQUIRED.
METERED AND UNMETERED CONDUCTORS
ALLOWED IN SAME CONDUIT

CUSTOMER WILL INSTALL
COMPANY FURNISHED METER
SOCKET. COMPANY WILL
FURNISH AND INSTALL METER

MAX. 5'-6"

MIN. 4'-0"
ABOVE GROUND
LINE

NEC APPROVED GROUNDING

GROUND LINE

Figure 23-2   Typical farm electric service, which is usually located
near the center of the electrical load. (Courtesy Potomac Edison
Power Co.)

CALCULATING SERVICE REQUIREMENTS

Farm services are calculated the same as any other building service; that is, the total loads are listed, a demand or diversity factor is applied, and the resulting figures indicate the service size. In sizing such a service, provisions should be made for future additions and loads. It is much less expensive to provide adequate service for these loads at the outset rather than having to enlarge the service entrance each time a new load is added. If a qualified consulting engineer is not available, then the local utility company will normally offer assistance to the farm and/or the electrical contractor in establishing a proper-sized service entrance.

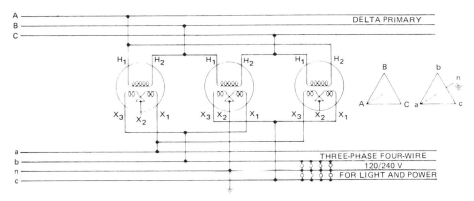

Figure 23-3 Transformer connection for a four-wire delta system.

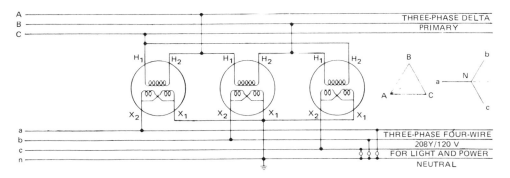

Figure 23-4 Transformer connection for a four-wire wye system.

LOCAL DISTRIBUTION

Since most farms utilize several barns, silos, and outbuildings, many feeders and subpanels are normally required for adequate electrical distribution. Here again, proper design and installation are necessary to keep voltage drop to the minimum and to provide a service within the budget of the farmer's means. See Fig. 23-5.

In general, two methods of installing feeder wires are currently in use: overhead and underground. Of the two, the latter is the type preferred although the initial installation is usually more expensive. Underground wiring, however, offers several advantages over overhead wiring, which usually offsets the additional initial cost. First, ice storms take their toll of overhead lines each year. The lines themselves are designed to take the load of ice on them, but an early snow or ice storm, before all leaves have left adjacent trees, can cause the tree limbs to break and fall across the lines, bringing them down

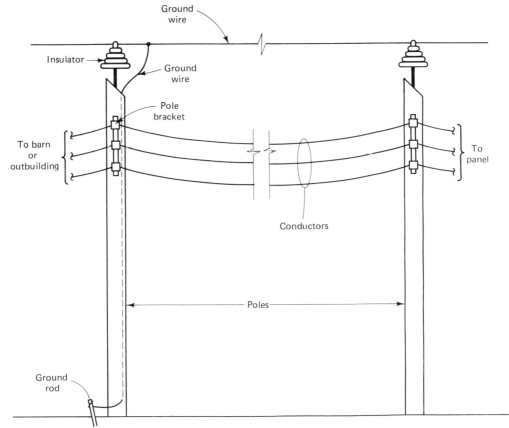

Figure 23-5   Typical farm overhead distribution system.

also. Underground wiring does not have this problem. Conductors buried in the ground are also safe from heavy farm equipment that might come into contact with overhead lines, and esthetic appearance is greatly improved with underground lines.

Underground electrical systems fall into two general categories from the standpoint of design and installation:

1. Direct burial, where the cables or conductors are buried directly in the earth.
2. Underground raceway systems, in which the conductors are installed at some time after the conduit or duct system is completed.

Direct-burial installations will range from small, single-conductor wires to multiconductor cables for power or communications or alarm systems. In any case, the conductors are installed in the ground either by placing them in an excavated trench, which is then backfilled, or by burying them directly by means of some form of cable plow, which opens a furrow, feeds the conductors into the furrow, and closes the furrow over the conductor.

Sometimes it becomes necessary to use lengths of conduit in conjunction with direct-burial installations, especially where the cables emerge on the surface of the ground or terminate at an outlet or junction box. Also, where the cables cross a roadway or concrete pavement, it is best to install a length of conduit under these areas in case the cable must be removed at a later date. By doing so, the road or concrete pad will not have to be disturbed.

The most popular type of raceway system is PVC rigid conduit. The size and number of conduits depends on the use of the system, number and size of conductors, and spare raceway capacity desired. The separation of the raceways depends on the mechanical strength of the system and the voltage that it will carry.

Figure 23-6 shows a cross section of a trench with direct-burial cable installed. Note the sand base on which the conductors lie to protect them from sharp stones and such. A treated board is placed over the conductors in the trench to offer protection during any digging that might occur in the future. Also, a continuous warning ribbon is laid in the trench, some distance above the board, to warn future diggers that electrical conductors are present in the area.

A cross section of a raceway system is shown in Fig. 23-7. Although the conductors are protected by the conduit, it is still recommended that a continuous warning ribbon be placed in the trench above the conduit should any digging be done in the area any time in the future.

Overhead electrical distribution systems are still used quite frequently around the farm. Wooden poles have proved to be an economical and satisfactory method of supporting overhead electrical lines in all areas. The overhead

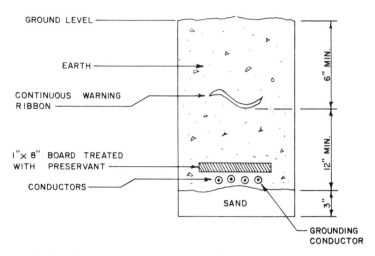

Figure 23-6   Cross section of a trench with direct-burial cable installed.

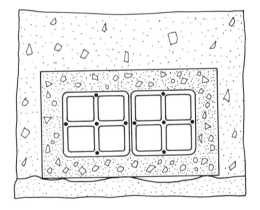

Figure 23-7   Cross section of a raceway system buried underground.

lines are strung from pole to pole using pin-type insulators, insulated devices, lightweight groups of suspension insulators, or similar means of support.

The height or length of poles depends primarily on their use and the terrain on which they are erected. Such items as voltage, number of circuits, changes in direction of the line, and length of conductor span are all important factors in determining the height of wooden poles for electrical distribution.

PREVENTING LIGHTNING DAMAGE

Each year many farmers and others who install and use electrical systems are faced with the problem of lightning damage, causing tremendous financial loss. In most cases, however, the destructive action of most lightning discharges can be quite effectively prevented by the use of properly installed lightning arrestors, proper grounding, and similar means of lightning protection.

Lightning is the discharge of enormous charges of static electricity accumulated on clouds. These charges are formed by the air currents striking the face of the clouds and causing condensation of the moisture in them. When wind strikes the cloud, these small particles of moisture are blown upward, carrying negative charges to the top of the cloud and leaving the bottom with positive charges. As very heavy rains or other forms of heavy condensation fall through a part of the cloud, one side becomes charged positively and the other side negatively, with many millions of volts' difference in potential.

When clouds under the condition just described come near enough to the ground or to another cloud with opposite charges, they will discharge to ground or to another cloud with explosive violence that everyone has seen from time to time.

Since there is a strong tendency for lightning discharges striking trees, structures, and other objects to travel on any metal parts that extend in the general direction of the discharge, lightning rods and properly grounded electrical systems can prevent much of the damage.

Lightning rods are made of copper or other material that is a good conductor of electricity, and a single lightning rod placed in the center of several objects will provide protection in a certain cone-shaped area around the objects. The diameter of the protective area is about four times the height of the rod. Therefore, a 30-foot lightning rod (above finished grade) would offer protection against lightning for approximately 60 feet in all directions from the base of the rod. Bear in mind, however, that the protection is only offered inside of the imaginary cone formed by the rod.

Overhead electrical lines running from building to building on the farm can be protected from lightning by running a bare ground wire above the current-carrying conductors. When this method is used, a good positive ground is provided to prevent lightning charges from entering the buildings.

Another means of protecting outdoor circuits and the equipment that they feed is by using lightning arrestors. In fact, some pump manufacturers are now offering built-in lightning arrestors in all their pumps. Arrestors are especially useful in preventing static electricity from coming in on the conductors feeding the pump motors.

Small, inexpensive lightning arrestors are available that are designed to prevent lightning surges that enter through the electrical wiring from damaging interior wiring and equipment. They have proved to immediately drain lightning surges harmlessly to ground. The arrestor is connected to the circuit by connecting the black wires to the hot wires of the system and the white wire to the grounded neutral or to a grounding electrode. The connection is usually the last switch box or disconnect controlling motors and other electrical equipment. For three-phase circuits, two arrestors of this type are required.

Variants of the previously mentioned arrestor designed for three-phase services up to 600 volts are offered by several manufacturers. One type has a porcelain housing and is designed for outdoor use only. Another type, with a metal enclosure, is suitable for both indoor and outdoor use and is available in one-, two-, and three-pole forms. These types also have better protective characteristics than the lower-voltage types and are therefore the type usually selected for protection of the larger motors used on the farm.

In general, the application of lightning arrestors will consist of the following:

1. Selecting the voltage rating of the arrestor to be used.
2. Choosing the types of arrestors needed.
3. Determining where the arrestors should be located to ensure adequate yet economical protection.

The first two items are relatively easy to determine, but determining the location of lightning arrestors should be given much consideration to ensure adequate protection.

The ideal location of the arrestor for motor installations is directly at the electrical terminals of the motor being protected. In some cases, however, it might be too costly or awkward to mount the arrestor at the motor itself. Furthermore, if the arrestors are mounted away from the terminals, say at the motor-control center, a single set of lightning arrestors may be used to protect two or more motors.

Every overhead power line feeding motors represents a possible source of destructive overvoltages caused by lightning coming in on the conductors. Therefore, lightning arrestors should be so applied that a voltage surge on these lines will be reduced to a value well below the impulse strength of all electrical equipment involved.

Other detailed requirements for lightning protection may be found in data supplied by the National Fire Protection Association, 470 Atlantic Ave., Boston, Massachusetts 02210.

FARM LOADS

Electrical wiring systems must be installed according to a careful plan that will ensure safety, adequacy, and convenience. The safety requirement will be satisfied by compliance with the NE Code, but the other two requirements can be satisfied only by careful planning of the wiring system.

The total electrical load on the farm consists of the lighting load, the miscellaneous load served by convenience outlets for various appliances and heaters, and the load driven by electric motors. The number of outlets and the approximate current used by each lamp, appliance, and motor are necessary for determining the number, size, and position of branch circuits and for laying out the entire wiring system.

The first step in determining the load is to make a sketch that shows the location of buildings and the distances between them. Such a sketch is recommended whether the entire wiring system is new, revamped, or a combination of the two. The inside dimensions of the buildings should be shown on the sketch, together with the major walls, doors, and partitions of each building. When necessary, additional sketches may be made of individual buildings to show the details.

The placement of all needed outlets can be conveniently indicated on a sketch. From the number and position of the outlets, the number of branch circuits can be planned. The sketch will also show what individual circuits are needed and where they should be installed.

LIGHTING

Proper lighting is an essential element of modern electrical living and working. The amount and type of lighting required should be determined by the various seeing tasks associated with the management of the home and farm. Types and locations of lighting outlets should conform with the lighting fixtures or equipment to be used. Unless a specified location is stated, lighting outlets can be located anywhere within the area under consideration to produce the desired lighting effects.

For wet and damp locations, such as certain farm buildings, use lamp receptacles with nonmetallic coverings, such as porcelain, plastic, or rubber. Lighting fixtures in feed-grinding rooms, feed-storage rooms, haymows, and other dusty locations should be of the dustproof type. See Chapter 14 for lighting calculations.

CONVENIENCE OUTLETS

Convenience outlets are intended to be used mainly for connecting portable devices to a source of electricity, and their location should be such that they are most convenient to the area in which the portable appliance is to be used. Here again careful planning can eliminate false economy by preventing hazards and inefficiency.

In farm buildings, especially those in which livestock are housed, convenience outlets should be located as high as can be reached conveniently in order to avoid damage by stock. Also, in general, the outlets should be mounted between studding or flush with the wall for additional protection.

Convenience outlets may be installed in the form of connectors suspended a convenient height above the floor on heavy-duty cord when it is desirable to keep appliance cords off the floor, as in cow stables or in areas where no convenient wall space is available. In buildings that house livestock and in which corrosive fumes are present, cords for suspension of connectors should be of such types as SJT, ST, SO, or SJO.

ELECTRIC-FENCE CONTROLLERS

Electric-fence controllers should be of the intermittent type approved by the National Board of Fire Underwriters. If approved, the approval will be shown on the nameplate or separate tags fastened to the controller. The current on the secondary, or output, side of the controller should be limited to 10 milli-amperes (mA) by the inherent characteristics of the controller. Tests have proved that an animal can safely stand 10 milliamperes of current for a very short time without suffering any ill effects.

From a safety standpoint, any controller that does not have an interrupter and current-limiting device as integral parts should not be used. The interrupter should be so designed that, if it fails to operate, it will fail in the open position and thus cut off the supply of electricity to the fence. Most controllers that meet the requirements of the authorities mentioned are also equipped with fuses to protect them against short circuits.

An electric fence should never be connected to the electric service either directly or through a light bulb or other resistance device without an approved controller. Many power suppliers will disconnect service if any one of these conditions is found. Even a current as small as 3 milliamperes, if not inter-rupted at regular intervals, can be fatal to humans as well as livestock. If you are ever called on to provide service for a fence controller, be sure an approved controller is used.

YARD LIGHTING

Yard lighting is a very important consideration for the farmstead. Properly located and installed lights can prevent many accidents on the farm, as well as greatly increase after-hours farm efficiency. The time required to do after-dark chores, such as watering, feeding, and housing livestock and storing machinery, can be greatly reduced with the aid of adequate light.

Outside lights should be so located that the most frequently used paths and work areas are well lighted. Outside feeding and watering areas should be illuminated by lights that can be controlled from the buildings most accessible to the areas. In the absence of buildings to support the lights, poles to which the lights can be attached should be installed.

The control of farm lighting should be analyzed in depth. In general, yard lighting should be controlled from both the house and outbuildings, either by means of three- and four-way switches or by relays. Low-voltage control systems can be installed with a master control panel located in the house to control every lighting circuit on the premises from one location, or at each individual location. Also, these low-voltage systems are advantageous when long distances are encountered between controls.

Whereas yard lighting consisted mainly of incandescent lamps a decade or so ago, these relatively inefficient light sources are now being replaced with quartz and HID lamps, all of which give off more lumens per watt. In any case, yard lighting around the farm is absolutely essential as a protection against thievery and predators and also as a safety factor to prevent accidents.

OUTLETS FOR VARIOUS FARM BUILDINGS

Each farm building has a specific purpose, and the wiring should be designed to be adequate for each application. The branch circuits terminate in outlet boxes wherever load will be applied. The outlets can generally be considered as lighting outlets, convenience outlets, and special outlets. Lighting outlets serve lamps; convenience outlets are provided for small appliances, plug-in equipment, and motor-driven devices rated at less than 3/4 horsepower; the special outlets are installed for motors that are larger than 3/4 horsepower and are permanently connected and for other large equipment, usually operating at 240 volts.

The characteristics of each specific load can usually be found on the nameplate of each piece of equipment, in accompanying operating instructions, or from the manufacturer. In each case, wire size and overcurrent protection should be selected to match the load. Also, due to the comparatively long runs encountered in farm wiring, voltage drop calculations should be made for each load and the wire size selected to hold the overall voltage drop down to 2 or 3 percent of the recommended voltage.

Lighting in most farm buildings has traditionally been of the incandescent type. However, the trend is now toward more efficient lighting practices, and fluorescent lighting fixtures should be used in these areas whenever possible.

Certain critical farm loads such as freezers, poultry heaters, milk coolers, a few emergency lights, and the like should be provided with a standby, emergency electrical system such as a gasoline-driven generator. Details of emergency power systems may be found in Chapter 25.

24

Wiring in Hazardous Locations

Articles 500 through 503 cover the requirements of electrical equipment and wiring for all voltages in locations where fire or explosion hazards may exist. Locations are classified depending on the properties of the flammable vapors, liquids, or gases or combustible dusts or fibers that may be present, as well as the likelihood that a flammable or combustible concentration or quality is present.

HAZARDOUS LOCATIONS

Any area in which the atmosphere or a material in the area is such that the arcing of operating electrical contacts, components, and equipment may cause an explosion or fire is considered as a hazardous location. In all such cases, explosion-proof equipment, raceways, and fittings are used to provide an explosion-proof wiring system.

Hazardous locations have been classified in the NE Code into certain class locations. Various atmospheric groups have been established on the basis of the explosive character of the atmosphere for the testing and approval of equipment for use in the various groups.

Class I Locations

Those locations in which flammable gases or vapors may be present in the air in quantities sufficient to produce explosive or ignitable mixtures are classified as Class I locations. Examples of such locations are interiors of paint spray booths where volatile, flammable solvents are used, inadequately ventilated pump rooms where flammable gas is pumped, and drying rooms for the evaporation of flammable solvents.

Class II Locations

Class II locations are those that are hazardous because of the presence of combustible dust. Class II, Division 1 locations are areas where combustible dust, under normal operating conditions, may be present in the air in quantities sufficient to produce explosive or ignitable mixtures; examples are working areas of grain-handling and storage plants and rooms containing grinders or pulverizers. Class II, Division 2 locations are areas where dangerous concentrations of suspended dust are not likely, but where dust accumulations might form.

Class III Locations

These locations are those areas that are hazardous because of the presence of easily ignitable fibers or flyings, but such fibers and flyings are not likely to be in suspension in the air in these locations in quantities sufficient to produce ignitable mixtures. Such locations usually include some parts of rayon, cotton, and textile mills, clothing manufacturing plants, and woodworking plants.

EXPLOSION-PROOF EQUIPMENT

The wide assortment of explosion-proof equipment now available makes it possible to provide adequate electrical installations under any of these hazardous conditions. However, the electrician must be thoroughly familiar with all NE Code requirements and know what fittings are available, how to install them properly, and where and when to use the various fittings.

A floor plan for a hazardous area is shown in Fig. 24-1. In general, either rigid metal conduit, IMC, or type MI cable is required for all hazardous locations, except for special flexible terminations and as otherwise permitted in the NE Code. The conduit must be threaded with a standard conduit cutting die that provides 3/4-inch taper per foot. The conduit should be made up wrench tight in order to minimize sparking in the event fault current flows through the raceway system (NE Code Article 500-1). Where it is impractical to make a threaded joint tight, a bonding jumper should be used. All boxes, fittings, and joints shall be threaded for connection to the conduit system and shall be an approved, explosion-proof type (Fig. 24-2). Threaded joints shall be made up with at least five threads fully engaged. Where it becomes necessary to employ flexible connectors at motor or fixture terminals (Fig. 24-3), flexible fittings approved for the particular class location shall be used.

Seal-off fittings (Fig. 24-4) are required in conduit systems to prevent the passage of gases, vapors, or flames from one portion of the electrical

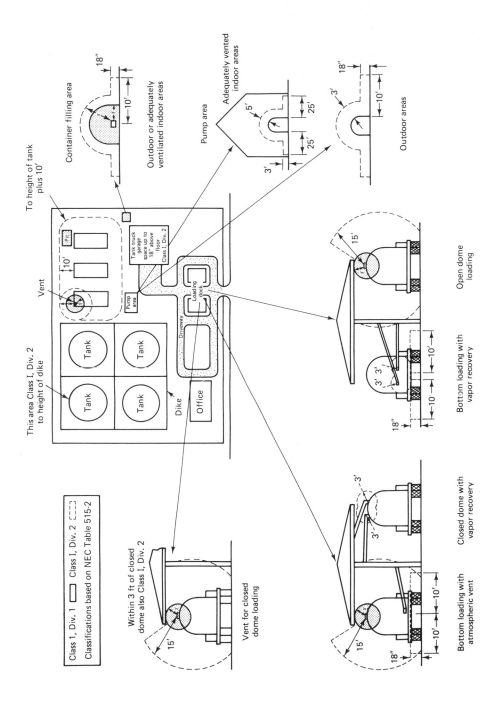

Class 1, Div. 1 ▭ Class I, Div. 2 ⊏⊐
Classifications based on NEC Table 515-2

Container filling area

Outdoor or adequately
ventilated indoor areas

Pump area

Adequately vented
indoor areas

Outdoor areas

To height of tank
plus 10'

This area Class I, Div. 2
to height of dike

Vent

Tank

Tank

Tank

Tank

Dike

Office

Driveway

Pump
area

Loading
dock

Tank truck
garage
space up to
18" above
floor
Class I, Div. 2

Within 3 ft of closed
dome also Class I, Div. 2

Vent for closed
dome loading

Open dome
loading

Bottom loading with
vapor recovery

Closed dome with
vapor recovery

Bottom loading with
atmospheric vent

Figure 24-1 Floor plan for a hazardous area showing the various fittings and wiring methods required. (Courtesy Crouse-Hinds.)

Groups C and D

| ECC | UNY Expansion | GR | GRU | GRUE | GRF |
| | U.S. Pat. 2,900,436 Can. Pat. 621,364 | | | | |

| GRJ | GRUO | GRUJ | GRU Union | CPU | GRJS |

Group D

| ELBD | | EJB | EXB |
| U.S. Pat. 3,020,332 | | | |

ER

Figure 24-2 Typical explosion-proof boxes used in hazardous locations. (Courtesy Appleton Electric Co.)

installation to another through the conduit. For Class I, Division 1 locations, the NE Code (Article 501-5) states that

> In each conduit run entering an enclosure for switches, circuit breakers, fuses, relays, resistors or other apparatus which may produce arcs, sparks or high temperatures, seals shall be placed as close as practicable and in no case more than 18 inches from such enclosures. There shall be no junction box or similar enclosure in the conduit run between the sealing fitting and the apparatus enclosure . . . .
>
> In each conduit run of 2-inch size or larger entering the enclosure or fitting housing terminals, splices or taps, and within 18 inches of such enclosure or fitting . . . .
>
> In each conduit run leaving the Class I, Division 1 hazardous area, the sealing fitting may be located on either side of the boundary of such hazardous area, but shall be so designed and installed that any gases or vapors which may enter the conduit system, within the Division 1 hazardous area, will not enter or be communicated to the conduit beyond the seal. There shall be no union, coupling, box or fitting in the conduit between the sealing fitting and the point at which the conduit leaves the Division 1 hazardous area . . . .

Sealing compound shall be approved for the purpose, shall not be affected by the surrounding atmosphere or liquids, and shall not have a melting point of less than 200°F (93°C). Most sealing-compound kits contain a powder in

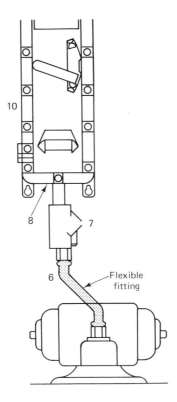

10

8   7

6   Flexible
fitting

Figure 24-3   Explosion-proof flexible fitting.

a polyethylene bag within an outer container. To mix, remove the bag of powder, fill the outside container, and pour in the powder and mix.

To pack the seal off, remove the threaded plug or plugs from the fitting and insert the asbestos fiber supplied with the packing kit. Tamp the fiber between the wires and the hub before pouring the sealing compound into the fitting. Then pour in the sealing cement and reset the threaded plug tightly. The fiber packing prevents the sealing compound (in the liquid state) from entering the conduit lines.

The seal-off fittings in Fig. 24-5 are typical of those used. The type in Fig. 24-5a is for vertical mounting and is provided with a threaded, plugged opening into which the sealing cement is poured. The seal off in Fig. 24-5b has an additional plugged opening in the lower hub to facilitate packing fiber around the conductors in order to form a dam for the sealing cement.

Most other explosion-proof fittings are provided with threaded hubs for securing the conduit as described previously. Typical fittings include switch and junction boxes, conduit bodies, union and connectors, flexible couplings, explosion-proof lighting fixtures, receptacles, and panelboard and motor starter enclosures. A practical representation of these and other fittings is shown in Fig. 24-6.

Figures 24-7 through 24-11 give examples of hazardous wiring installations in Class I and Class II locations.

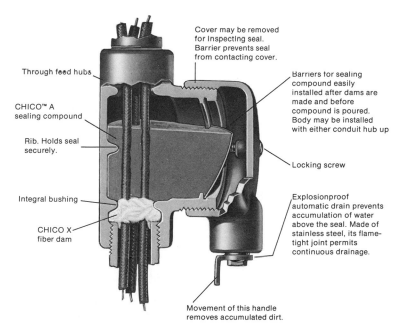

Figure 24-4   Seal-off fittings are required in conduit systems to prevent passage of gases, vapors, or flames from one portion of an electrical system to another. (Courtesy Crouse-Hinds.)

GARAGES AND SIMILAR LOCATIONS

Garages and similar locations where volatile or flammable liquids are handled or used as fuel in self-propelled vehicles (including automobiles, buses, trucks, and tractors) are not usually considered critically hazardous locations. However, the entire area up to a level 18 inches above the floor is considered a Class I, Division 2 location, and certain precautionary measures are required by the NE Code. Likewise, any pit or depression below floor level shall be considered a Class I, Division 2 location, and the pit or depression may be judged as Class I, Division 1 location if it is unvented.

Normal raceway (conduit) and wiring may be used for the wiring method above this hazardous level, except where conditions indicate that the area concerned is more hazardous than usual. In this case, the applicable type of explosion-proof wiring may be required.

Approved seal-off fittings should be used on all conduit passing from hazardous area to nonhazardous area. The requirements set forth in NE Code Sections 501-5 and 501-5(b)(2) shall apply to horizontal as well as vertical boundaries of the defined hazardous areas. Raceways embedded in a masonry floor or buried beneath a floor are considered to be within the hazardous area above the floor if any connections or extensions lead into or through such an area.

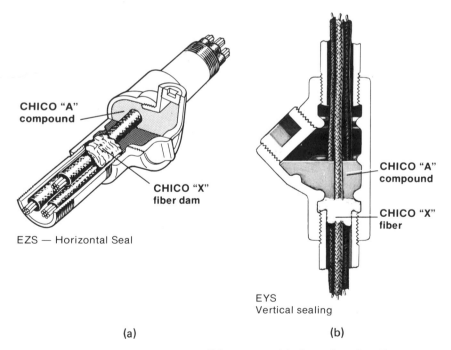

CHICO "A" compound

CHICO "X" fiber dam

EZS — Horizontal Seal

CHICO "A" compound

CHICO "X" fiber

EYS
Vertical sealing

(a)   (b)

Figure 24-5 Typical seal-off fittings used in hazardous locations.
(Courtesy Crouse-Hinds.)

Figure 24-12 shows a typical automotive service station with applicable NE Code requirements. Note that space in the immediate vicinity of the gasoline-dispensing island is denoted as Class I, Division 1, to a height of 4 feet above grade. The surrounding area, within a radius of 20 feet of the island, falls under Class I, Division 2, to a height of 18 inches above grade. Bulk storage plants for gasoline are subject to comparable restrictions.

AIRPORT HANGARS

Buildings used for storing or servicing aircraft in which gasoline, jet fuels, or other volatile flammable liquids or gases are used fall under Article 513 of the NE Code. In general, any pit or depression below the level of the hangar floor is considered to be a Class I, Division 1 location. The entire area of the hangar including any adjacent and communicating area not suitably cut off from the hangar is considered to be a Class I, Division 2 location up to a level of 18 inches above the floor. The area within 5 feet horizontally from aircraft power plants, fuel tanks, or structures containing fuel is considered to be a Class I, Division 2 hazardous location; this area extends upward from the floor to a level 5 feet above the upper surface of wings and of engine enclosures.

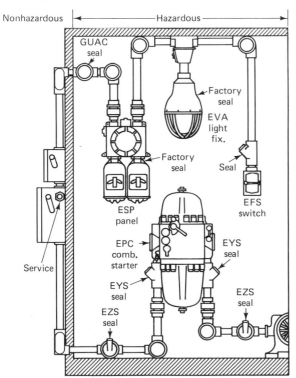

Figure 24-6   A practical representation of explosion-proof fittings used in various classes of locations. (Courtesy Crouse-Hinds.)

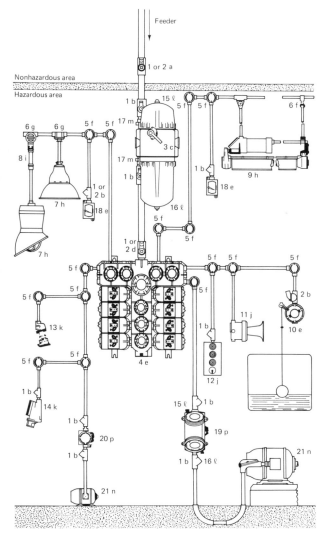

Key to Numerals

1　Sealing fitting. EYS for horizontal or vertical.
2　Sealing fitting. EZS for vertical or horizontal conduits.
3　Circuit breaker. Type EPC.
4　Panelboard. EDP. Branch circuits are factory sealed. No seals required in mains or branches unless 2 inches or over in size.
5　Junction box. Series GUA, GUB, GUJ have threaded covers. Series CPS has ground flat surface covers.
6　Fixture hanger. EFHC, GUAC, or EFH.
7　Lighting fixture. EV Series.
8　Flexible fixture support. ECHF or EFH.
9　Fluorescent fixture. EVF.
10　Float switch. EMS.
11　Signal. ETH horns and sirens. ERS bells.
12　VISULARM® EKP.
13　Plug receptacle. CES delayed action.
14　Plug receptacle. FSQ. Interlocked with switch.
15　Breather. ECD.
16　Drain. ECD.
17　Union. UNY.
18　Switch. Series EFS.
19　Manual line starter. FLF.
20　Manual line starter. GUSC.
21　Motors, explosion proof.

National Electrical Code References

a　Sec. 501-5(a)(4). Seal required where conduit passes from hazardous to nonhazardous area.
b　Sec. 501-5(a)(1). Seals required within 18 inches of all arcing devices.
c　Sec. 384-16. Circuit breaker protection required ahead of panelboard.
d　Sec. 501-5(a)(1). Seals required if conduit is 2 inches or larger.
e　Sec. 501-6(a). All arcing devices must be explosion proof.
f　Sec. 501-4(a). All boxes must be explosion proof and threaded for rigid or IMC conduit.
g　Sec. 501-9(a)(4). All boxes and fittings for support of lighting fixtures must be approved for Class I locations.
h　Sec. 501-9(a)(1). All lighting fixtures fixed or portable must be explosion proof.
i　Sec. 501-9(a)(3). Pendant fixture stems must be threaded rigid or IMC conduit. Conduit stems if over 12 inches must have flexible connector, or must be braced.
j　Sec. 501-14(a). All signal and alarm equipment irrespective of voltage must be approved for Class I, Division I locations.
k　Sec. 501-12. Receptacles and plugs must be explosion proof and provide grounding connections for portable equipment.
ℓ　Sec. 501-5(f). Breathers and drains needed in all humid locations.
m　Sec. 501-4(a). All joints and fittings must be explosion proof.
n　Sec. 501-8(a). Motor must be suitable for Class I.
p　Art. 430. Motor overcurrent protection.

Figure 24-7　　Class I, Division 1 electrical installation. (Courtesy Crouse-Hinds.)

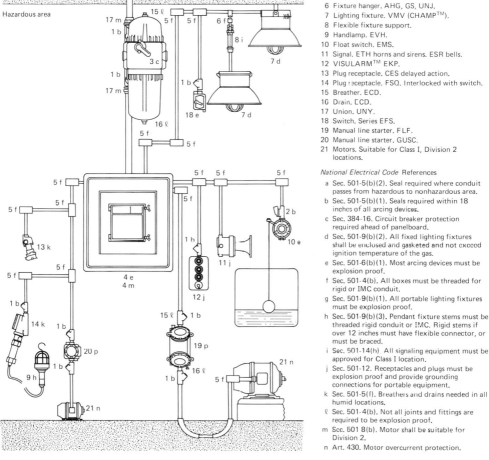

Feeder

Nonhazardous area

Hazardous area

Key to Numerals

1 Sealing fitting. EYS for horizontal or vertical.
2 Sealing fitting. EZS for vertical or horizontal conduits.
3 Circuit breaker. Type EPC.
4 Panelboard. D2BP, N2PB. Branch circuits are factory sealed.
5 Junction box or conduit fitting. NJB. OBROUND®.
6 Fixture hanger. AHG, GS, UNJ.
7 Lighting fixture. VMV (CHAMP™).
8 Flexible fixture support.
9 Handlamp. EVH.
10 Float switch. EMS.
11 Signal. ETH horns and sirens. ESR bells.
12 VISULARM™ EKP.
13 Plug receptacle. CES delayed action.
14 Plug receptacle. FSQ. Interlocked with switch.
15 Breather. ECD.
16 Drain. ECD.
17 Union. UNY.
18 Switch. Series EFS.
19 Manual line starter. FLF.
20 Manual line starter. GUSC.
21 Motors. Suitable for Class I, Division 2 locations.

National Electrical Code References

a Sec. 501-5(b)(2). Seal required where conduit passes from hazardous to nonhazardous area.
b Sec. 501-5(b)(1). Seals required within 18 inches of all arcing devices.
c Sec. 384-16. Circuit breaker protection required ahead of panelboard.
d Sec. 501-9(b)(2). All fixed lighting fixtures shall be enclosed and gasketed and not exceed ignition temperature of the gas.
e Sec. 501-6(b)(1). Most arcing devices must be explosion proof.
f Sec. 501-4(b). All boxes must be threaded for rigid or IMC conduit.
g Sec. 501-9(b)(1). All portable lighting fixtures must be explosion proof.
h Sec. 501-9(b)(3). Pendant fixture stems must be threaded rigid conduit or IMC. Rigid stems if over 12 inches must have flexible connector, or must be braced.
i Sec. 501-14(h) All signaling equipment must be approved for Class I location.
j Sec. 501-12. Receptacles and plugs must be explosion proof and provide grounding connections for portable equipment.
k Sec. 501-5(f). Breathers and drains needed in all humid locations.
ℓ Sec. 501-4(b). Not all joints and fittings are required to be explosion proof.
m Sec. 501-8(b). Motor shall be suitable for Division 2.
n Art. 430. Motor overcurrent protection.

Figure 24-8   Electrical diagram for Class I, Division 2, power and lighting. (Courtesy Crouse-Hinds.)

Adjacent areas in which hazardous vapors are not likely to be released, such as stock rooms and electrical control rooms, should not be classed as hazardous when they are adequately ventilated and effectively cut off from the hangar itself by walls or partitions. All fixed wiring in a hangar not within a hazardous area as defined in Section 513-2 must be installed in metallic raceways or shall be Type MI or Type ALS cable; the only exception is wiring in nonhazardous locations as defined in Section 513-2(d), which may be of any type recognized in Chapter 3 of this code. Figure 24-13 summarizes the NE Code requirements for airport hangars.

THEATERS

The NE Code recognizes that hazards to life and property due to fire and panic exist in theaters, cinemas, and the like. The NE Code therefore requires certain precautions in these areas in addition to those for commercial installations. These requirements include the following:

1. Proper wiring of motion picture projection rooms (Article 540).
2. Heat-resistant, insulated conductors for certain lighting equipment [Article 520-43(b)].
3. Adequate guarding and protection of the stage switchboard and proper control and overcurrent protection of circuits (Article 520-22).
4. Proper type and wiring of lighting dimmers [Articles 520-52(e) and 520-25].
5. Use of proper types of receptacles and flexible cables for stage lighting equipment (Article 520-45).
6. Proper stage flue damper control (Article 520-49).
7. Proper dressing-room wiring and control (Articles 520-71, 72, and 73).
8. Fireproof projection rooms with automatic projector port closures, ventilating equipment, emergency lighting, guarded work lights, and proper location of related equipment (Article 540).

Outdoor or drive-in motion picture theaters do not present the inherent hazards of enclosed auditoriums. However, the projection rooms must be properly ventilated and wired for the protection of the operating personnel.

HOSPITALS

Hospitals and other health-care facilities fall under Article 517 of the NE Code. Part B of Article 517 covers the general wiring systems of health-care facilities. Part C covers essential electrical systems for hospitals. Part D gives

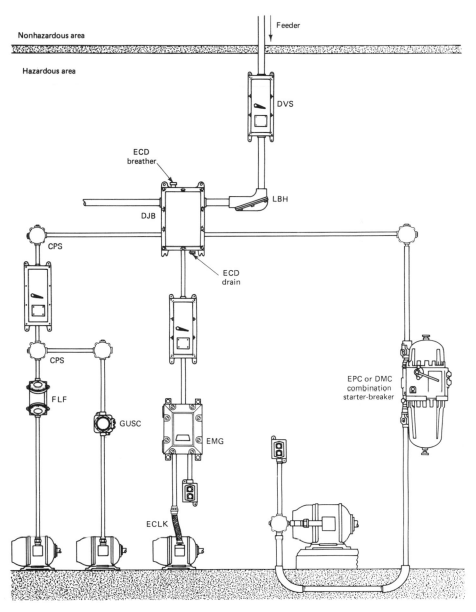

Figure 24-9 Diagram for Class II power installation. (Courtesy Crouse-Hinds.)

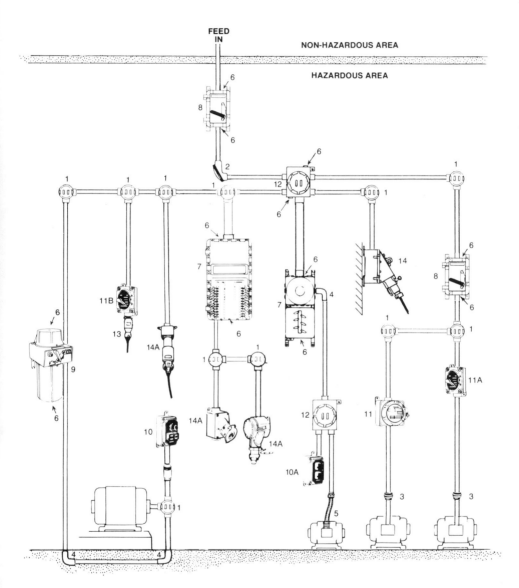

Figure 24-10   Power diagram for Class II, Division 1 installation.
(Courtesy Appleton Electric Co.)

CLASS II, DIVISION 1

NEC 502-1. GENERAL.
The general rules of this Code shall apply to the installation of electrical wiring and apparatus in locations classified as Class II under Section 500-5 except as modified by this Article.

NEC 502-2. TRANSFORMERS AND CAPACITORS.
Not a product of Appleton.

NEC 502-3. SURGE PROTECTION, CLASS II, DIVISIONS 1 AND 2.
Not a product of Appleton.

NEC 502-4. WIRING METHODS.
1—CAST CONDUIT OUTLET BOX UNILETS (Section J).
2—PULLING UNILET (Section J).
3—UNIONS (Section I).
4—ELBOWS, PLUGS AND MISCELLANEOUS FITTINGS (Section I).
5—FLEXIBLE COUPLINGS (Section I).

NEC 502-5. SEALING, CLASS II, DIVISIONS 1 AND 2.
6—DRAIN AND BREATHERS (Section I).

NEC 502-6. SWITCHES, CIRCUIT BREAKERS, MOTOR CONTROLLERS, AND FUSES.
7—PANELBOARDS (Section P). Also refer to NEC Art. 430.
8—CIRCUIT BREAKER UNILETS (Section O). Also refer to NEC Art. 430.
9—COMBINATION CIRCUIT BREAKERS AND MOTOR STARTERS (Section O). Refer also to NEC Art. 430.
10—COMBINATION PUSH BUTTON CONTROL STATION AND PILOT LIGHT (Section M). Refer also to NEC Art. 430.
10A—PUSH BUTTON CONTROL STATION (Section M). Refer also to NEC Art. 430.
11—MOTOR STARTER OR SWITCH (GUSC) UNILET (Sections N and O). Refer also to NEC Art. 430.
11A—MOTOR STARTER (EFD) UNILET* (Section O). Refer also to NEC Art. 430.
11B—SWITCH UNILET* (Section N).

NEC 502-7. CONTROL TRANSFORMERS AND RESISTORS.
12—DUST-IGNITION-PROOF UNILET ENCLOSURES (Section K).

NEC 502-8. MOTORS AND GENERATORS.
Not a standard product listing, but motors provided with some Reelite® products.

NEC 502-9. VENTILATING PIPING.
Not a product of Appleton.

NEC 502-10. UTILIZATION EQUIPMENT, FIXED AND PORTABLE.
UTILIZATION EQUIPMENT—Products such as portable lamps must be approved for Class II locations.
13—CORD CONNECTORS (Section L). The Appleton ECC is an approved method for permanent connection from a rigid conduit system to a portable appliance. Supplies power to utilization equipment.

NEC 502-11. LIGHTING FIXTURES.
Example diagram on next page.

NEC 502-12. FLEXIBLE CORDS, CLASS II, DIVISIONS 1 AND 2.
Not a standard product listing, but flexible cord is provided with Reelite® products designed for hazardous locations (Section W3).

NEC 502-13. RECEPTACLES AND ATTACHMENT PLUGS, CLASS II, DIVISIONS 1 AND 2.
14—PLUGS AND RECEPTACLES with interlocking safety switch (Section L).
14A—PLUGS AND RECEPTACLES (Section L).

NEC 502-14. SIGNALING, ALARM, REMOTE-CONTROL AND COMMUNICATION SYSTEMS.

NEC 502-15. LIVE PARTS, CLASS II, DIVISIONS 1 AND 2.
There shall be no exposed live parts.

NEC 502-16. GROUNDING, CLASS II, DIVISIONS 1 AND 2.
The use of Appleton Static Discharge Reels (Section Z) is recommended for use with metal parts of equipment such as frames, motors, machinery, and other generators of electricity. See pages 65 and 74 of this booklet.

Figure 24-10   continued

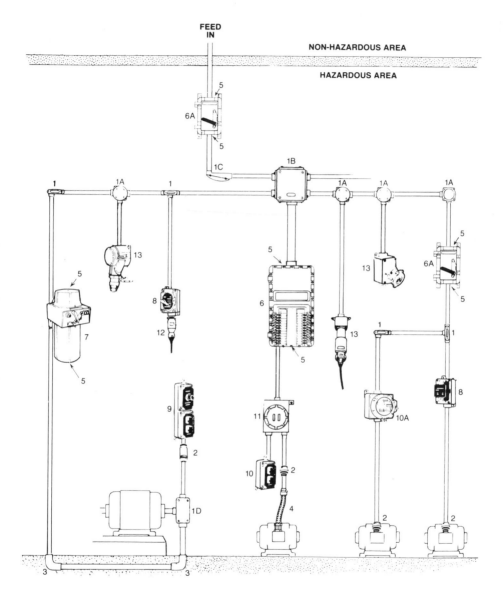

Figure 24-11 Power diagram for Class II, Division 2 installation. (Courtesy Appleton Electric Co.)

CLASS II, DIVISION 2

NEC 502-1. GENERAL
The general rules of this Code shall apply to the installation of electrical wiring and apparatus in locations classified as Class II under Section 500-5 except as modified by this Article.

NEC 502-2. TRANSFORMERS AND CAPACITORS.
Not a product of Appleton.

NEC 502-3. SURGE PROTECTION, CLASS II, DIVISIONS 1 and 2.
Not a product of Appleton.

NEC 502-4. WIRING METHODS.
1—CAST CONDUIT OUTLET BODIES (Section A).
1A—CAST CONDUIT OUTLET BOX UNILETS® (Section A).
1B—RS JUNCTION UNLET (Section C).
1C—PULLING UNILET (Section A).
1D—FS AND FD JUNCTION UNILET WITH BLANK CAST COVER (Section B).
2—UNIONS (Section I).
3—ELBOWS, PLUGS AND MISCELLANEOUS FITTINGS (Section I).
4—FLEXIBLE COUPLINGS (Section I).

NEC 502-5. SEALING, CLASS II, DIVISIONS 1 AND 2.
5—DRAIN AND BREATHERS (Section I).

NEC 502-6. SWITCHES, CIRCUIT BREAKERS, MOTOR CONTROLLERS AND FUSES.
6—PANELBOARDS (Section P). Also refer to NEC Art. 430.
6A—CIRCUIT BREAKER UNILETS (Section O). Also refer to NEC Art. 430.
7—COMBINATION CIRCUIT BREAKERS AND MOTOR STARTERS (Section O). Refer also to NEC Art. 430.
8—SWITCH UNILETS (Section N). Also refer to NEC Art. 430.
9—COMBINATION PUSH BUTTON, CONTROL STATION AND PILOT LIGHT (Section M). Refer also to NEC Art. 430.
10—PUSH BUTTON CONTROL STATION (Section M). Refer also to NEC Art. 430.
10A—MOTOR STARTER OR SWITCH (GUSC) UNILET® (Sections N and O). Refer also to NEC Art. 430.

NEC 502-7. CONTROL TRANSFORMERS AND RESISTORS.
11—DUST-IGNITION-PROOF UNILET ENCLOSURES (Section K).

NEC 502-8. MOTORS AND GENERATORS.
Not a standard product listing, but motors provided with some Reelite® products (Section W3).

NEC 502-9. VENTILATING PIPING.
Not a product of Appleton.

NEC 502-10. UTILIZATION EQUIPMENT, FIXED AND PORTABLE.
UTILIZATION EQUIPMENT. Products such as portable lamps must be approved for Class II locations.
12—CORD CONNECTORS (Section L). The Appleton ECC is an approved method for permanent connection from a rigid conduit system to a portable appliance.Supplies power to utilization equipment.

NEC 502-11. LIGHTING FIXTURES.
Best described in diagram on next page.

NEC 502-12. FLEXIBLE CORDS, CLASS II, DIVISIONS 1 AND 2.
Not a standard product listing, but flexible cord is provided with Reelite® products designed for hazardous locations (Section W3).

NEC 502-13. RECEPTACLES AND ATTACHMENT PLUGS, CLASS II, DIVISIONS 1 AND 2.
13—PLUGS AND RECEPTACLES (Section L).

NEC 502-14. SIGNALING, ALARM, REMOTE-CONTROL AND COMMUNICATION SYSTEMS.

NEC 502-15. LIVE PARTS, CLASS II, DIVISIONS 1 AND 2.
There shall be no exposed live parts.

NEC 502-16. GROUNDING, CLASS II, DIVISIONS 1 AND 2.
The use of Appleton Static Discharge Reels (Section Z) is recommended for use with metal parts of equipment such as frames, motors, machinery, and other generators of electricity. See pages 65 and 74 of this booklet.

Figure 24-11   continued

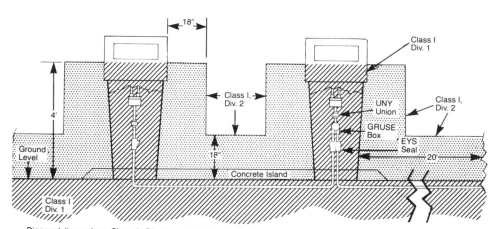

Diagonal lines show Class I, Div. 1 spaces. Formerly, NEC classified the space "within 18" extending horizontally from the dispenser up to 4 feet from its base" as Class I, Div. 1. With the 1981

Code, this space is now classified Class I, Div. 2. Space within the dispenser and below grade level remains as Class I, Div. 1. Note that per NEC 514-6 seals are the first fittings installed after the conduit

emerges from the concrete and that seals are installed in each conduit entering the dispensers.

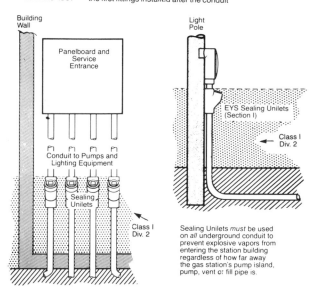

Sealing Unilets *must* be used on *all* underground conduit to prevent explosive vapors from entering the station building regardless of how far away the gas station's pump island, pump, vent or fill pipe is.

Figure 24-12   Typical wiring installation of a service station. (Courtesy Appleton Electric Co.)

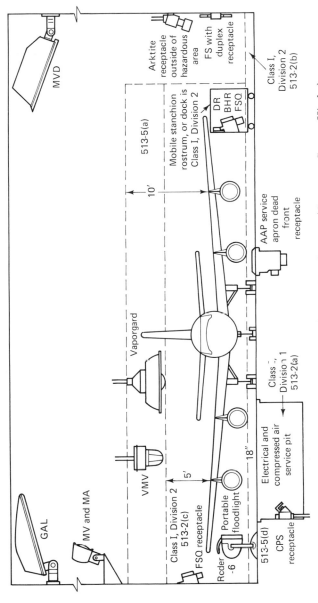

Figure 24-13 Wiring requirement for airplane hangar. (Courtesy Crouse-Hinds.)

the performance criteria and wiring methods to minimize shock hazards to patients in electrically susceptible patient areas. Part E covers the requirements for electrical wiring and equipment used in inhalation anesthetizing locations.

With the widespread use of x-ray equipment of varying types in health-care facilities, electricians are often required to wire and connect equipment such as discussed in Article 660 of the NE Code. Conventional wiring methods are used, but provisions should be made for 50- and 60-ampere receptacles for medical x-ray equipment (Section 660-3b).

Anesthetizing locations of hospitals are deemed to be Class I, Division 1, to a height of 5 feet above floor level. Gas storage rooms are designated as Class I, Division 1, throughout.

The NE Code recommends that wherever possible electrical equipment for hazardous locations be located in less hazardous areas. It also suggests that by adequate, positive-pressure ventilation from a clean source of outside air the hazards may be reduced or hazardous locations limited or eliminated. In many cases the installation of dust-collecting systems can greatly reduce the hazards in a Class II area.

25

Standby and Emergency Power Systems

Essential service loads, such as hospitals, most public buildings, and many kinds of industrial operations that cannot tolerate a power outage require two separate sources of power. These sources, of course, should not be paralleled. Rather, an automatic transfer system is generally used and is built in many different designs and various voltages.

The most important factor in selecting a standby system is size and/or capacity. A careful study should be made to determine exactly what degree of service it is desirable to maintain during a power outage. How much and what kind of equipment must the system operate? What is the total wattage?

It is advisable at this stage to adopt a long-range view since it often becomes necessary to add more equipment to the load in future years. It may be desirable to provide full protection at the outset, since the higher cost for a larger unit is offset by special circuit wiring costs necessary for selected protection. As needs grow, the set may still be made adequate by restricting protection to essential equipment. The decision on whether to provide full or selected protection will most often be decided by the particular operation for which the unit is provided. A hospital, for example, requires full protection for all essential equipment, while another building may only wish to maintain emergency lighting.

Another factor to consider at this stage is the altitude of the locale and the effect it may have on the rating of the standby set. If the point of installation is at a higher altitude than that of the manufacturer, the unit is generally derated 4 percent per 1000 feet above sea level.

Also, complex factors, such as surge currents of electric motors, the magnetizing current of transformers, feedback from elevator loads, heating effects, and electromagnetic interference from solid-state loads, must be thoroughly explored and understood before system parameters are finalized.

STANDBY SYSTEM PARAMETERS

Fuel

Before the engine can be selected, careful consideration must be given to selecting the proper type of fuel. Often, the type of fuel is one of the most important considerations in final engine selection.

Fuel options are gasoline, liquid propane (LP) gas, natural gas, and diesel. The specific advantages of each are discussed later. However, several general points that may outrule all others should be considered at the outset:

1. Availability of the various fuels in your particular location.
2. Local regulations governing the storage of gasoline and gaseous fuels.

Cooling

Air and liquid cooling systems are available. Liquid cooling may be either a radiator or freshwater cooling system. The determining factor between air or water cooling will generally be the size of the unit. Air cooling must be confined to smaller units up through 15 kilowatts. In these smaller sizes it is possible to force air over the engine and generator at a sufficient rate to cool it properly and thus take advantage of lower equipment and maintenance costs. Larger units, above 15 kilowatts, should as a rule be equipped with a liquid system for maximum cooling protection.

Regardless of what capacity or cooling system is used, the location, size, and temperature of the room must be considered. City water cooling, remote radiators, and heat exchangers are available for special applications.

Load Transfer Switch

Electric power companies require transfer switches and usually supervise their installation, since power lines on which utility personnel work are affected. There are two types of transfer switches: manual and automatic. The manual transfer switch must be operated by a person on duty either at the plant or a remote station. The plant is started by a manual switch, and the load is transferred from one source to another by a hand-operated double-throw switch.

An automatic transfer switch starts the plant and transfers the load automatically, not requiring the attention of an operator. With automatic transfer switches, the power outage can be limited to less than 10 seconds. An automatic load transfer switch is almost a necessity for applications where uninterrupted power is of prime importance, such as a hospital, or where equipment is remotely located or where health or safety is at stake.

TYPES OF FUEL

There is a wide range of electricity generating sets from 1 to 85 kilowatts, using gasoline and/or gaseous fuel. The advantages of these two types of fuel are discussed next. It should be remembered that each type of fuel is superior in some respects and less favorable in others. The following general comments should be considered in the light of individual preferences, local conditions, and fuel availability.

ADVANTAGES OF GAS OPERATION

1. *More efficient operation:*   Because this fuel is already a gas, it mixes readily with the air, and combustion is more complete.
2. *Longer life:*   There is no lead in gas, so deposits do not accumulate. The engine runs cleaner and lasts longer, requiring less maintenance and fewer oil changes.
3. *Quick starting:*   Starting after long shutdowns is quicker with natural or LP gas because, unlike gasoline, gas remains "fresh" in storage. Both gasoline and gas engines are usually easier to start under difficult conditions and in extremely low temperatures.
4. *Lower initial cost:*   A gasoline or gas set generally costs less than diesel.

BTU Content Is Important

There are several different gaseous fuel mixtures: natural, manufactured, and bottle gas, or, as it is known commercially, butane or propane. Care should be taken to check the Btu content of gaseous fuels. The Btu content must be at least 1100 to get the same output as a similar gasoline set. Butane and propane gases meet this requirement. Some natural or manufactured gases, however, run as low as 450 Btu. A system using 450-Btu fuel must be derated 40 to 50 percent. For further information on gaseous fuel operation, write to the Onan factory for Technical Bulletin No. T015 or Manual SP-1020, 1400 73rd Avenue, N.E., Minneapolis, Minnesota 55432.

ADVANTAGES OF DIESEL OPERATION

Diesel-driven electricity generating sets are available in capacities ranging from 3 to 750 kilowatts. Improved diesel engine design has made diesel fuel a reliable and dependable power source for most applications. The advantages of diesel operation are as follows:

1. *Costs less to operate:*   Diesel fuel costs less per gallon and has a higher BTU content than gasoline. Fuel consumption is also much less as

compared to gasoline. This efficiency and saving in fuel costs increase with the size of the unit.

2. *Reduced maintenance costs:*   The heavier weight construction inherent in diesel design, plus the absence of points, plugs, and condensers, reduces service and repair needs by almost 50 percent over gasoline.

3. *Safe, low-volatility fuel:*   The high flash point and low volatility of diesel fuel reduce the possibility of fire or explosion from fumes and leakage, making underground storage unnecessary.

4. *Quick response:*   Diesels start promptly on diesel fuel alone without auxiliary devices, reach operating speed quickly, and handle lugging loads easily.

PLANNING AND SELECTION OF EQUIPMENT

When an emergency generating set is being considered, there is a tendency to skimp and cut corners on equipment on the premise that the system may never be used. Equipment that cannot handle the job is little better than none at all. When standby power is needed, it is desperately needed, and the best and most reliable equipment pays off at just such times.

26

Basis of Electrical Blueprint Reading

An electrical blueprint is an exact copy or reproduction of an original drawing, consisting of lines, symbols, dimensions, and notations to accurately convey an engineer's design to the workers who install the electrical system on the job. A blueprint, therefore, is an abbreviated language for conveying a large amount of exact, detailed information, which would otherwise take many pages of manuscript or hours of verbal instruction to convey.

In every branch of electrical work, there is often occasion to read an electrical drawing. Electricians, for example, who are responsible for installing the electrical system in a new building usually consult an electrical drawing to locate the various outlets, the routing of circuits, the location and size of panelboards, and other similar electrical details in preparing a bid. The electrical estimator of a contracting firm must refer to electrical drawings to determine the quality of material needed. Electricians in industrial plants consult schematic diagrams when wiring electrical controls for machinery. Plant maintenance personnel use electrical blueprints in troubleshooting. Circuits may be tested and checked against the original drawings to help locate any faulty points in the installation.

The ideal electrical drawing should show in a clear, concise manner exactly what is required of the workers. The amount of data shown on such a drawing should be sufficient, but not overdone. This means that a complete set of electrical drawings could consist of only one 8-1/2- by 11-inch sheet, or it could consist of several dozen sheets, depending on the complexity of the given project.

ELECTRICAL SYMBOLS

In preparing electrical drawings, most engineers and designers use symbols adopted by the United States of America Standards Institute (USASI). However, many designers frequently modify these standard symbols to suit their own needs. For this reason, most drawings will have a symbol list or legend. Figure 26-1 shows a list of electrical symbols currently being used by one consulting-engineering firm.

It is evident from this list that many of the symbols have the same basic form, but their meanings are different because of some slight difference in the symbol. For example, all the outlet symbols in Fig. 26-2 have the same basic form, a circle; however, the addition of a line or a dot to the circle gives each an individual meaning. It is also apparent that the difference in meaning may be indicated by the addition of letters or an abbreviation to the symbol. Therefore, a good procedure to follow in learning the different symbols is to first learn the basic form and then apply the variations for obtaining different meanings.

Some of the symbols used on electrical drawings are abbreviations, such as WP for weatherproof and AFF for above finished floor. Others are simplified pictographs, such as ⌾ for a double floodlight fixture or ▭ for an infrared electric heater with two quartz lamps.

In some cases, the symbols are combinations of abbreviation and pictographs, such as [F] for fusible safety switch, [DT]$_{30}$ for a double-throw safety switch, and [N]$_{60}$ for a nonfusible safety switch. In each example, a pictograph of a switch enclosure has been combined with an abbreviation, F (fusible), DT (double throw), and N (nonfusible), respectively. The numerals indicate the bus-bar capacity in amperes.

The lighting-outlet symbols represent both incandescent and fluorescent types; a circle usually represents an incandescent fixture and a rectangle, a fluorescent one. All these symbols are designed to indicate the physical shape of a particular fixture and should be drawn as close to scale as possible.

The type of mounting used for all lighting fixtures is usually indicated in a lighting-fixture schedule, which is shown on the drawings or in the written specification.

The type of lighting fixture is identified by a numeral placed inside a triangle near each lighting fixture. A complete description of the fixture identified by the symbols must be given in the lighting-fixture schedule and should include the manufacturer, catalog number, number and type of lamps, voltage, finish, mounting, and any other information needed for a proper installation of the fixtures.

A single-pole switch is used to control lighting from one point, while three- and four-way switches are used in combination to control a single light or a group of lights from two or more points.

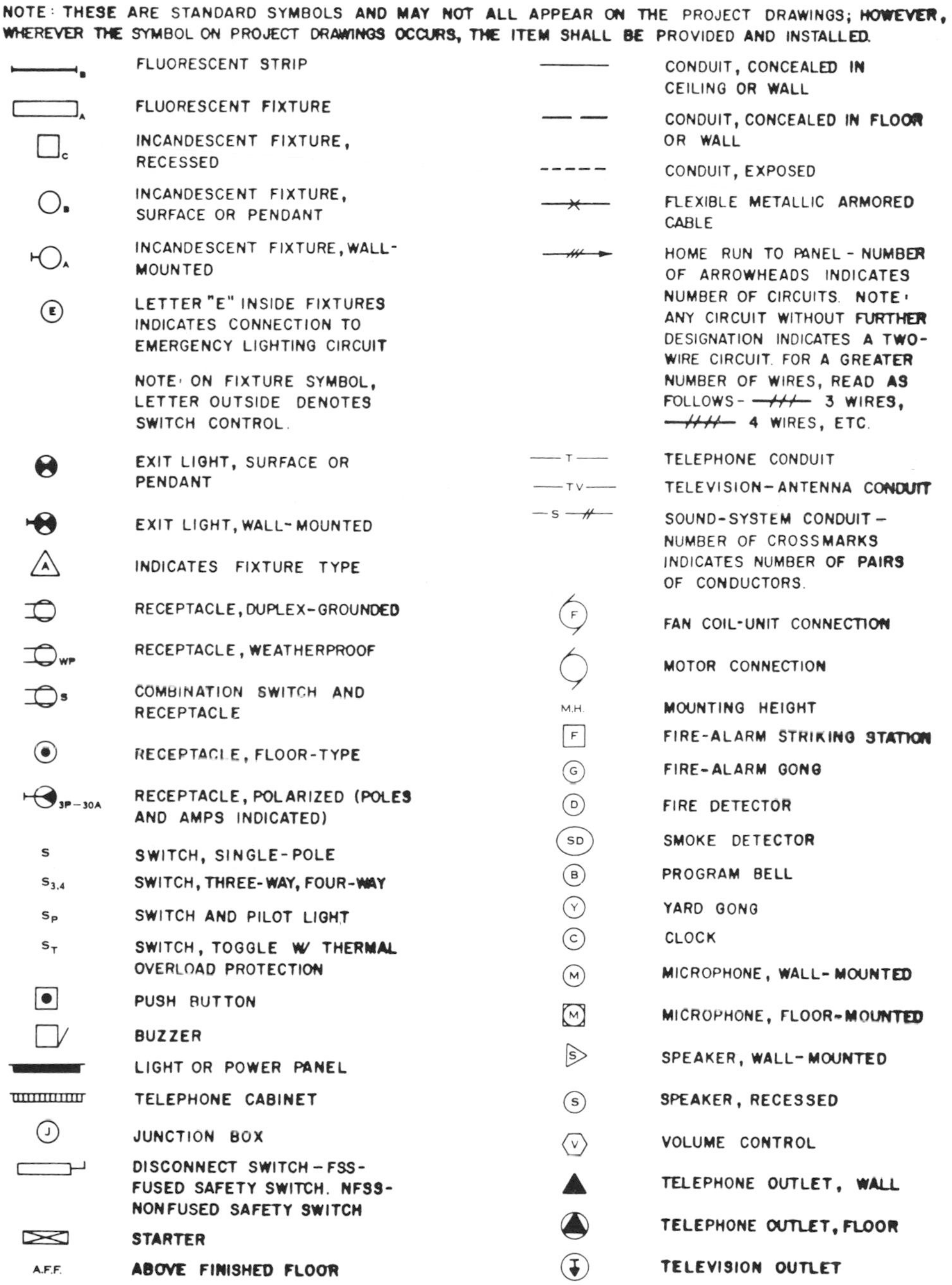

NOTE: THESE ARE STANDARD SYMBOLS AND MAY NOT ALL APPEAR ON THE PROJECT DRAWINGS; HOWEVER, WHEREVER THE SYMBOL ON PROJECT DRAWINGS OCCURS, THE ITEM SHALL BE PROVIDED AND INSTALLED.

FLUORESCENT STRIP

FLUORESCENT FIXTURE

INCANDESCENT FIXTURE, RECESSED

INCANDESCENT FIXTURE, SURFACE OR PENDANT

INCANDESCENT FIXTURE, WALL-MOUNTED

LETTER "E" INSIDE FIXTURES INDICATES CONNECTION TO EMERGENCY LIGHTING CIRCUIT

NOTE: ON FIXTURE SYMBOL, LETTER OUTSIDE DENOTES SWITCH CONTROL.

EXIT LIGHT, SURFACE OR PENDANT

EXIT LIGHT, WALL-MOUNTED

INDICATES FIXTURE TYPE

RECEPTACLE, DUPLEX-GROUNDED

RECEPTACLE, WEATHERPROOF

COMBINATION SWITCH AND RECEPTACLE

RECEPTACLE, FLOOR-TYPE

RECEPTACLE, POLARIZED (POLES AND AMPS INDICATED)

SWITCH, SINGLE-POLE

SWITCH, THREE-WAY, FOUR-WAY

SWITCH AND PILOT LIGHT

SWITCH, TOGGLE W/ THERMAL OVERLOAD PROTECTION

PUSH BUTTON

BUZZER

LIGHT OR POWER PANEL

TELEPHONE CABINET

JUNCTION BOX

DISCONNECT SWITCH – FSS-FUSED SAFETY SWITCH. NFSS-NONFUSED SAFETY SWITCH

STARTER

A.F.F. ABOVE FINISHED FLOOR

CONDUIT, CONCEALED IN CEILING OR WALL

CONDUIT, CONCEALED IN FLOOR OR WALL

CONDUIT, EXPOSED

FLEXIBLE METALLIC ARMORED CABLE

HOME RUN TO PANEL - NUMBER OF ARROWHEADS INDICATES NUMBER OF CIRCUITS. NOTE: ANY CIRCUIT WITHOUT FURTHER DESIGNATION INDICATES A TWO-WIRE CIRCUIT. FOR A GREATER NUMBER OF WIRES, READ AS FOLLOWS - —///— 3 WIRES, —///— 4 WIRES, ETC.

—T— TELEPHONE CONDUIT

—TV— TELEVISION—ANTENNA CONDUIT

—S—//— SOUND-SYSTEM CONDUIT – NUMBER OF CROSSMARKS INDICATES NUMBER OF PAIRS OF CONDUCTORS.

F FAN COIL-UNIT CONNECTION

MOTOR CONNECTION

M.H. MOUNTING HEIGHT

F FIRE-ALARM STRIKING STATION

G FIRE-ALARM GONG

D FIRE DETECTOR

SD SMOKE DETECTOR

B PROGRAM BELL

Y YARD GONG

C CLOCK

M MICROPHONE, WALL-MOUNTED

M MICROPHONE, FLOOR-MOUNTED

S SPEAKER, WALL-MOUNTED

S SPEAKER, RECESSED

V VOLUME CONTROL

TELEPHONE OUTLET, WALL

TELEPHONE OUTLET, FLOOR

TELEVISION OUTLET

Figure 26-1   Electrical symbols used by a consulting engineering firm.

A two-pole switch is used to control a series of lights on two separate circuits with only one motion or to control single-phase 240-volt loads. The switch and pilot-light combination are used where it is practical, if the lighting fixture controlled by the switch is located in a closet, basement, or attic space.

Door switches in residential construction are commonly used to control closet lights. The operation of these switches is very simple: when the door is closed, the bottom of the switch (located in the door jamb) is depressed, which opens the circuit; when the door is opened, the switch button is released, closing the circuit, and the light comes on.

Main distribution centers, panelboards, transformers, safety switches, and other similar electrical components are indicated by electrical symbols on floor plans and by a combination of symbols and semipictorial drawings in riser diagrams.

A detailed description of the service equipment is usually given in the panelboard schedule or in the written specifications. However, on small projects, the service equipment is sometimes indicated only by notes on the drawings.

Circuit and feeder wiring symbols have been nearly standardized. Most circuits concealed in the ceiling or wall are indicated by a solid line; a broken line is used for circuits concealed in the floor or ceiling below; and exposed raceways are indicated by short dashes.

The number of conductors in a conduit or raceway system may be indicated in the panelboard schedule under the appropriate column, or the information may be shown on the floor plan.

Symbols for communication and signal-systems, as well as symbols for light and power, are drawn to an appropriate scale and accurately located with respect to the building; this reduces the number of references made to the architectural drawings. Where extremely accurate final location of outlets and equipment is necessary, exact dimensions are given on larger-scale drawings and shown on the plans.

Each different category in a signal system is usually represented by a distinguishing basic symbol. Every item of equipment or outlet in that

DUPLEX
RECEPTACLE

COMBINATION SWITCH
& RECEPTACLE

DUPLEX RECEPTACLE WITH
WATERPROOF COVER

RECEPTACLE,
FLOOR TYPE

Figure 26-2   Examples of various receptacle symbols.

category of the system is identified by that basic symbol. To further identify items of equipment or outlets in the category, a numeral or other identifying mark is placed within the open basic symbol. In addition, all such individual symbols used on the drawings should be included in the symbol list or legend.

ANALYZING A SET OF WORKING DRAWINGS

The most practical way to learn how construction documents are prepared is to analyze an existing drawing prepared by an architectural firm.

The first page of a set of blueprints is called the *title sheet* (see Fig. 26-3). Formats will vary from firm to firm, but in general this first sheet contains the name and location of the building project, the name of the architectural firm, an index to the drawings, and usually the legend or key for materials used in the drawing sheets to follow.

The second page in the set usually includes a drawing of the plot on which the building or buildings are to be constructed. A breakdown of soil borings taken to examine the bearing qualities of the earth may also be included.

A drawing of a typical plot plan is shown in Fig. 26-4. This plan is drawn to scale with pertinent dimensions and indicates the location of the building on the site as well as its walks and drives. Natural ground slope of the side is indicated by grade lines. The drawing also shows what the finish grade of the site is to be when the building is erected and certain earth moving has been accomplished.

The elevation figures on the plot plan in Fig. 26-4 are based on a *bench mark,* which is a permanent object in the area with a predetermined elevation in feet above sea level. In all other drawings in the set that follow, the first-floor level of the building is given a base elevation of 100 feet for the purpose of easier dimensioning. Then other elevations are figured from that. The plot plan also contains indications of existing buildings, streets, trees, and permanent landmarks.

On smaller projects, utility services such as water, electricity, telephone, cable TV, and gas and sewer lines are also indicated on the plot plan. On larger projects, a separate sheet titled "Utility Plot Plan" will be included to facilitate the reading of the drawings.

Drawing sheets immediately following the plot plan will normally contain the building floor plans, examples of which appear in Fig. 26-5. Floor plans, more so than any other type of drawing, show how the building is laid out and are the most readable to the nonprofessional.

A floor plan represents a cut horizontally through a building at approximately eye level and appears on the sheet as though the top half of the cut is removed and the viewer is looking down on the building from above. A separate drawing is made for each floor or level (including basement). Shown

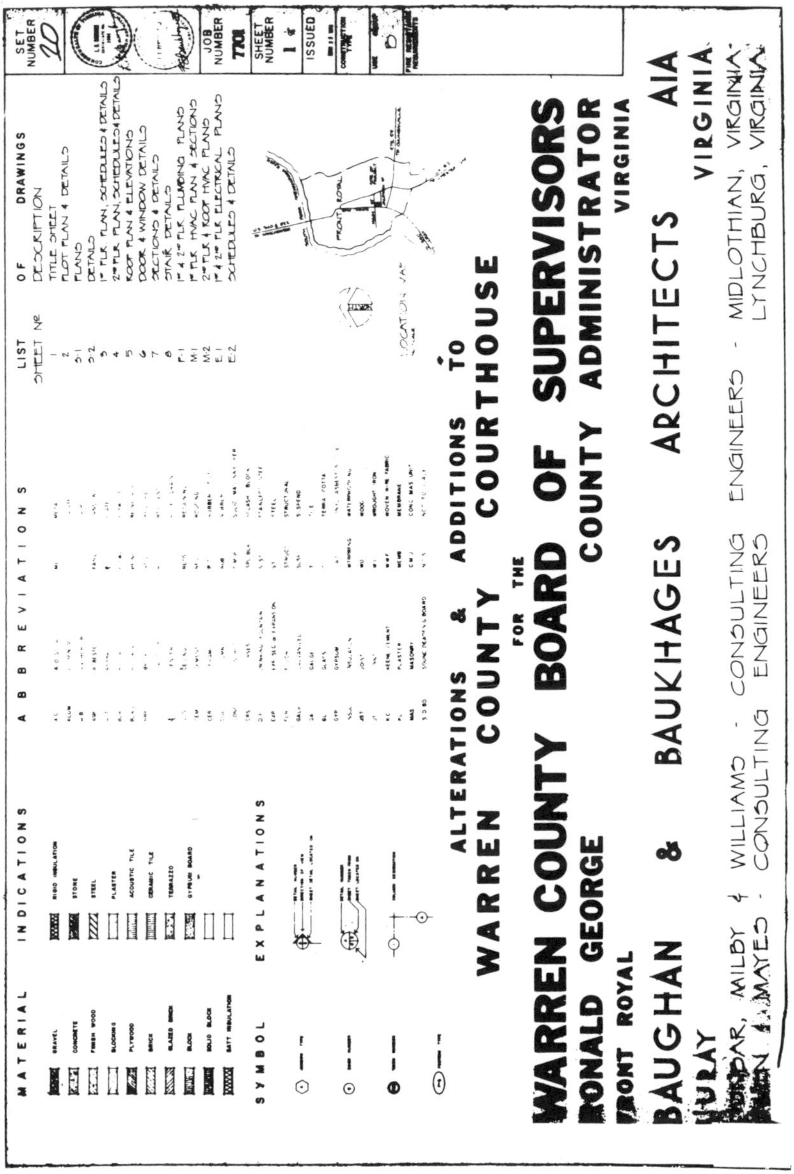

Figure 26-3 The first page of a set of construction drawings, normally called the "Title Sheet."

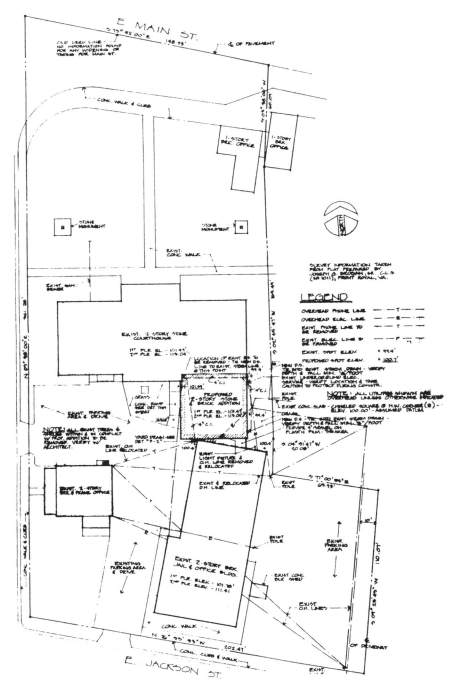

Figure 26-4   Typical plot plan.

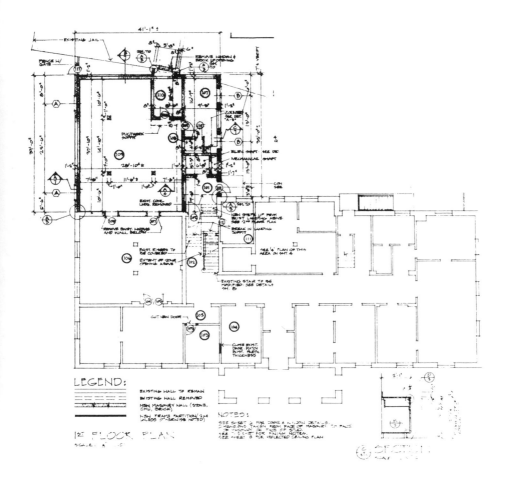

Figure 26-5   Floor plan drawing.

in the floor plans are the location and arrangement of all walls, partitions, doors, windows, and stairways, with indication of dimensions and materials used. Floor plans also contain many references to elevation, section, and other plan drawings on other sheets and could be called the key sheets of a set of working drawings.

Following the floor plans are drawing sheets showing the building elevations. Elevations are head-on, vertical views of a building or wall area in a single plane. Therefore, the front elevation of a building is as if the viewer looked straight at the building from the front, the right side elevation is as if the viewer looked straight at the right side of the building, and so forth.

Elevations also show the materials with which the walls are constructed, that is, glass, concrete block, and so on. Symbols are used to indicate most of the finishes although notes are sometimes employed. Figure 26-6 shows examples of elevation drawings.

Drawing sheets showing various cross sections appear in Fig. 26-7. These cross sections appear as if the area of the building in question is sliced open to reveal construction details that cannot be shown by drawings of elevations or plans. Both small- and large-scale sections are necessary throughout a set of blueprints to enable workers to construct the building. A thorough study of cross sections and other views is a great aid in learning how a building is constructed, since great detail is shown.

The parts of the working drawing discussed thus far show most of the construction details. However, certain construction conditions cannot be shown adequately in the scale in which most drawings are made. Therefore, larger-scaled drawings with greater detail must sometimes be made of some areas to ensure that the workers will understand what is to be done. These are termed special, or large-scale, detail drawings and are found throughout a set of working drawings to better explain construction and design features.

Structural Drawings

Following the architectural drawings, usually designated by sheet numbers A-1, A-2, and so on, are structural drawings showing details of footings, foundation, structural framing, and other structural details.

Footings are concrete "feet" placed in the ground and sometimes reinforced with steel bars on which the foundation and the subsequent building load are placed. Soil conditions, together with the weight of the building, determine the size, design, and number of footings. Building weight is measured in terms of *dead* and *live* loads. Dead load is the stationary weight of the building itself and the permanently fixed equipment. Live load is made up of movable equipment in the building and the humans who use it.

Foundation walls serve as a base on which the building is built and carry the load of the building to the footings and earth below. These walls must be strong enough to resist the side pressure of the earth. Steel reinforcing

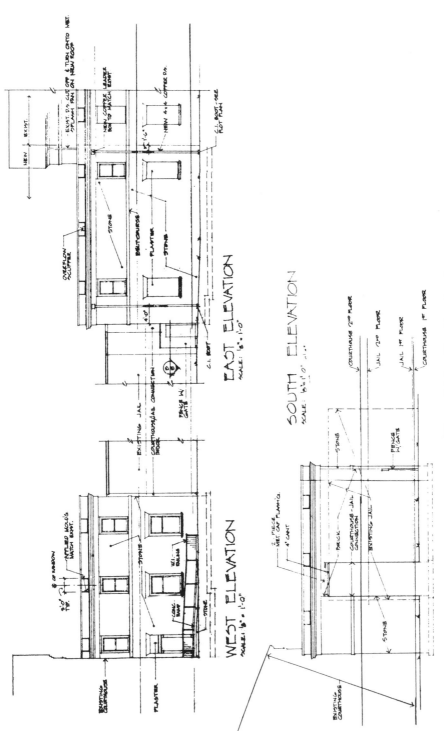

Figure 26-6   Elevation drawings.

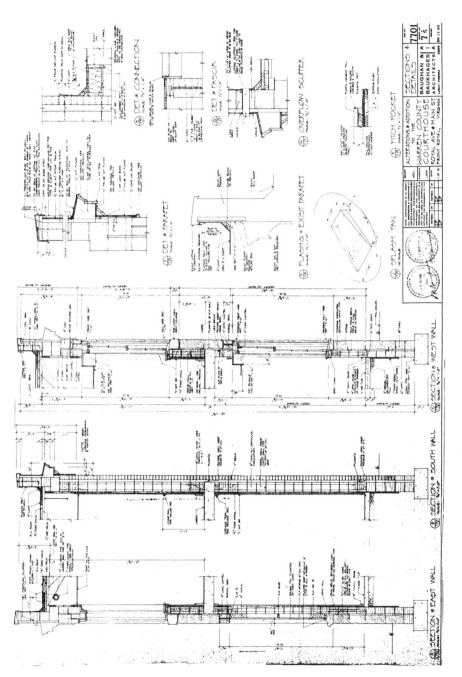

Figure 26-7 Cross sections of a building.

bars are used in foundation walls, which are deep into the ground to offset the extra heavy earth pressure. Another function of the walls is to keep moisture out of the underground parts of the building. Usually a waterproofing compound is applied to the exterior surface of foundation walls.

Figures 26-8, 26-9, and 26-10 show examples of structural drawings used on actual installations. Note the details, schedules, and other pertinent data included.

Mechanical Drawings

Mechanical drawings are usually included in all sets of working drawings for projects of any consequence. These drawings include details for the plumbing heating, ventilating, air-conditioning, and temperature control systems.

Most architects hire consulting engineers who specialize in this field to design and prepare working drawings and specifications for this portion of the project. In doing so the design must be coordinated very carefully with the architect as well as the electrical designer to ensure smooth integration of the proper facilities with the building design.

Mechanical drawings are highly diagrammatical and are used to locate pipes, fixtures, ductwork, equipment, and so forth. Detailed descriptions of this and other equipment are found in the written specifications. Figure 26-11 shows a typical plumbing drawing and Fig. 26-12 shows a heating and cooling system layout.

Electrical Drawings

Electrical drawings are prepared in much the same way as mechanical drawings; that is, architects hire consulting engineers to design the system and then prepare working drawings and written specifications.

Electrical drawings prepared by consulting engineers are unique drawings. A complete set of working drawings for an electrical system will usually consist of the following:

1. A plot plan showing the location of the building on the property and all outside electrical wiring, including the service entrance. This plan is drawn to scale with the exception of the various electrical symbols, which must be enlarged to be readable.
2. Floor plans showing the walls and partitions for each floor level. The physical locations of all wiring and outlets are shown for lighting, power, signal and communication, special electrical systems, and related electrical equipment. Again, the building partitions are drawn to scale, as are such electrical items as fluorescent lighting fixtures, panelboards, and

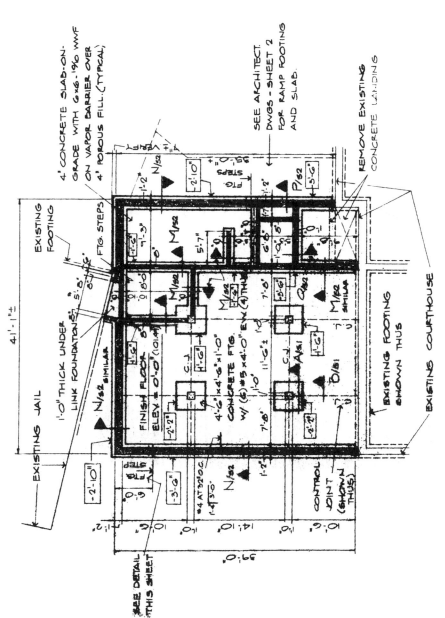

Figure 26-8   Structural drawing showing foundation plan.

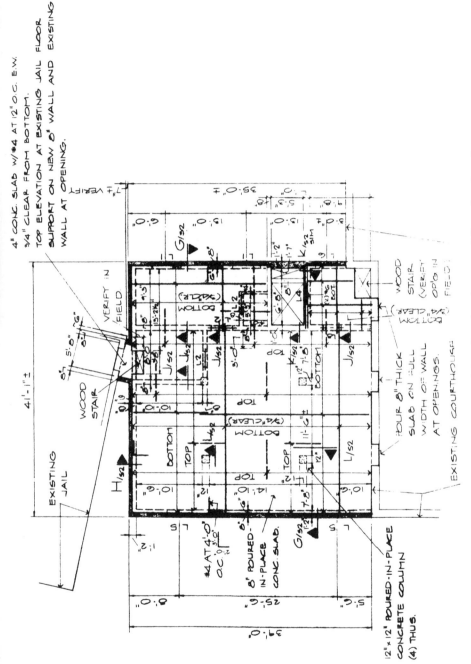

Figure 26-9   Structural drawing showing second-floor framing plan.

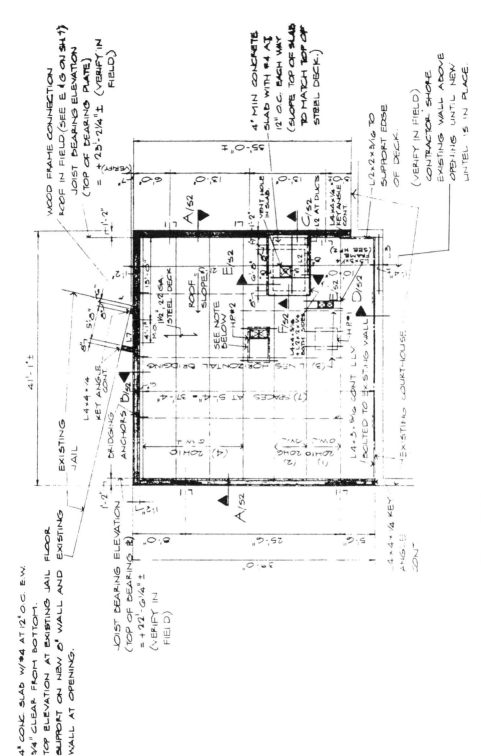

Figure 26-10 Structural drawing showing roof framing plan.

313

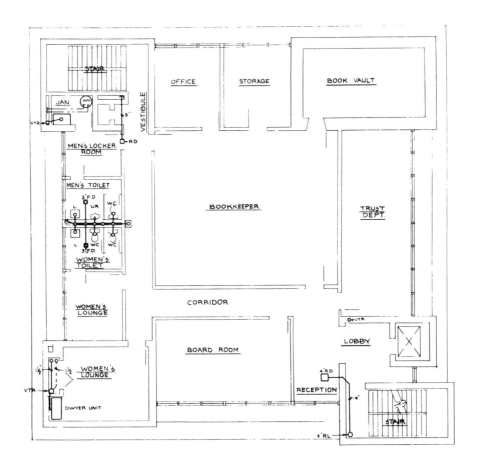

Figure 26-11   Typical plumbing drawing.

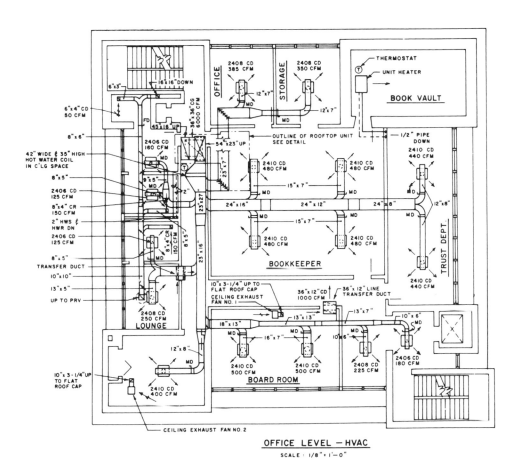

Figure 26-12   Heating and cooling system layout.

switchgear. The locations of other electrical outlets and similar components are only approximated on the drawings because they have to be
exaggerated.

3. Power-riser diagrams to show the service-entrance and panelboard
components.

4. Control wiring schematic diagrams.

5. Schedules, notes, and large-scale details on construction drawings.

To be able to "read" electrical as well as other types of drawings, one
must become familiar with the meaning of symbols, lines, and abbreviations
used on the drawings and learn how to interpret the message conveyed by
the drawings.

An orthographic-projection drawing is the most frequently used type of
drawing for electrical systems in building construction.

The electrical floor plan of a building shows all outside walls, interior
partitions, windows, doors, and the like, along with the location of all electrical
outlets, panelboards, branch circuits, feeders, service wire and equipment, and
the other details necessary for making a correct installation.

The lighting layout for a building will usually be laid out on a separate
floor plan from the power wiring. Electrical symbols are used almost exclusively in showing the location of electrical outlets on building floor plans.

ELECTRICAL SPECIFICATIONS

The electrical specifications for a building or project are the written descriptions of work and duties required of the owner, architect, and engineer.
Together with the working drawings, these specifications form the basis of
the contract requirements for the construction.

Those who prepare the construction drawings and specifications must
always be on the alert for conflicts between the working drawings and the
written specifications. Such conflicts occur particularly when:

1. Architects or engineers use standard or prototype specifications and
attempt to apply them to specific working drawings.

2. Previously prepared standard drawings are to be changed or amended
by reference in the specifications only; the drawings themselves are not
changed.

In such situations, the conflicts must be cleared up, preferably before
the work is undertaken, to avoid added cost either to the owner, architect,
or contractor performing the work.

In general, the electrical specifications should show the grade of materials
to be used on the project and the manner in which the electrical system shall
be installed. A typical electrical specification follows.

DIVISION 16—ELECTRICAL

Section 16A—General Provisions

1. The "Instructions to Bidders," "General Conditions," and "General Requirements" of the architectural specifications govern work under this Section.
2. It is understood and agreed that the Electrical Contractor has, by careful examination of the Plans and Specifications, and the site where appropriate, satisfied himself as to the nature and location of the work, and all conditions which must be met in order to carry out the work under this Section of the contract.
3. Scope of the Work
   a. The scope of the work consists of the furnishing and installing of complete electrical systems—exterior and interior—including miscellaneous systems. The Electrical Contractor shall provide all supervision, labor, materials, equipment, machinery, and any and all other items necessary to complete the systems. The Electrical Contractor shall note that all items of equipment are specified in the singular; however, the Contractor shall provide and install the number of items of equipment as indicated on the drawings and as required for complete systems.
   b. It is the intention of the Specifications and Drawings to call for finished work, tested, and ready for operation.
   c. Any apparatus, appliance, material, or work not shown on drawings but mentioned in the specifications, or vice versa, or any incidental accessories necessary to make the work complete and perfect in all respects and ready for operation, even if not particularly specified, shall be furnished, delivered, and installed by the Contractor without additional expense to the Owner.
   d. Minor details not usually shown or specified, but necessary for proper installation and operation, shall be included in the Contractor's estimate, the same as if herein specified or shown.
   e. With submission of bid, the Electrical Contractor shall give written notice to the Architect of any materials or apparatus believed inadequate or unsuitable, in violation of laws, ordinances, rules; and any necessary items or work omitted. In the absence of such written notice, it is mutually agreed that the Contractor has included the cost of all required items in his proposal, and that he will be responsible for the approved satisfactory functioning of the entire system without extra compensation.
4. Electrical Drawings
   a. The Electrical drawings are diagrammatic and indicate the general arrangement of fixtures, equipment and work included in the contract. Consult the Architectural drawings and details for exact location of fixtures and equipment; where same are not definitely located, obtain this information from the Architect.
   b. Contractor shall follow drawings in laying out work and check drawings of other trades to verify spaces in which work will be installed. Maintain maximum headroom and space conditions at all points. Where headroom or space conditions appear inadequate, the Architect shall be notified before proceeding with installation.

      c. If directed by the Architect, the Contractor shall, without extra charge, make reasonable modifications in the layout as needed to prevent conflict with work of other trades or for proper execution of the work.

5. Codes, Permits, and Fees

      a. Contractor shall give all necessary notices, including electric and telephone utilities, obtain all permits and pay all government taxes, fees and other costs, including utility connections or extensions in connection with his work; file all necessary plans, prepare all documents and obtain all necessary approvals of all governmental departments having jurisdiction; obtain all required certificates of inspection for his work and deliver same to the Architect before request for acceptance and final payment for the work.

      b. Contractor shall include in the work, without extra cost to the Owner, any labor materials, services, apparatus, drawings (in addition to contract drawings and documents) in order to comply with all applicable laws, ordinances, rules and regulations, whether or not shown on drawings and/or specified.

      c. Work and materials shall conform to the latest rules of the National Board of Fire Underwriters' Code, Regulations of the State Fire Marshal, and with applicable local codes and with all prevailing rules and regulations pertaining to adequate protection and/or guarding of all moving parts, or otherwise hazardous conditions. Nothing in these specifications shall be construed to permit work not conforming to the most stringent of applicable codes.

      d. The National Electric Code, the Local Electric Code, and the electrical requirements as established by the State and Local Fire Marshal, and rules and regulations of the power company serving the project, are hereby made part of the drawings or specifications to make the work comply with these requirements. The Electrical Contractor shall so notify the Architect.

6. Shop Drawings

      a. The Electrical Contractor shall submit five (5) copies of the shop drawings to the Architect for approval within thirty (30) days after the award of the general contract. If such a schedule cannot be met, the Electrical Contractor may request in writing for an extension of time to the Architect. If the Electrical Contractor does not submit shop drawings in the prescribed time, the Architect has the right to select the equipment.

      b. Shop drawings shall be submitted on all major pieces of electrical equipment, including service entrance equipment, lighting fixtures, panel boards, switches, wiring devices and plates and equipment for miscellaneous systems. Each item of equipment proposed shall be a standard catalog product of an established manufacturer. The shop drawing shall give complete information on the proposed equipment. Each item of the shop drawings shall be properly labeled, indicating the intended service of the material, the job name and Electrical Contractor's name.

      c. The shop drawings shall be neatly bound in five (5) sets and submitted to the Architect with a letter of transmittal. The letter of transmittal shall list each item submitted along with the manufacturer's name.

      d. Approval rendered on shop drawings shall not be considered as a guarantee of measurements or building conditions. Where drawings are approved, said approval does not mean that drawings have been checked in detail, said approval does not in any way relieve the Contractor from his responsibility, or necessity of furnishing material or performing work as required by the contract drawings and specifications.

7. Cooperation with Other Trades
   a. The Electrical Contractor shall give full cooperation to other trades and shall furnish (in writing, with copies to Architect) any information necessary to permit the work of all trades to be installed satisfactorily and with least possible interference or delay.
   b. Where the work of the Electrical Contractor will be installed in close proximity to work of other trades, or where there is evidence that work of the Electrical Contractor will interfere with the work of other trades, he shall assist in working out space conditions to make a satisfactory adjustment. If so directed by the Architect, the Electrical Contractor shall prepare composite working drawings and sections at a suitable scale clearly showing how his work is to be installed in relation to the work of other trades, he shall make necessary changes in his work to correct the condition without extra charge.
   c. The complexity of equipment and the variation between equipment manufacturers requires complete coordination of all trades. The Contractor who offers, for consideration, substitutes of equal products of reliable manufacturers, has to be responsible for all changes that affect his installation and the installation equipment of other trades. All systems and their associated controls must be completely installed, connected, and operating to the satisfaction of the Architect prior to final acceptance and contract payment.

8. Temporary Electrical Service
   a. The Electrical Contractor shall be responsible for all arrangements and costs for providing at the site temporary electrical metering, main switches and distribution panels as required for construction purposes. The distribution panels shall be located at a central point designated by the Architect; the General Contractor shall indicate prior to installation whether three-phase or single-phase service is required.

9. Electrical Connections
   a. The Electrical Contractor shall provide and install power wiring to all motors and electrical equipment complete and ready for operation including disconnect switches and fuses. Starters, relays and accessories shall be furnished by others unless otherwise noted, but shall be installed by the Electrical Contractor. This Contractor shall be responsible for checking the shop drawings of the equipment manufacturer to obtain the exact location of the electrical rough-in and connections for equipment installed.
   b. The Mechanical Contractor will furnish and install all temperature control wiring and all interlock wiring unless otherwise noted.
   c. It shall be the responsibility of the Electrical Contractor to check all motors for proper rotation.

10. As-built Drawings
    a. The Electrical Contractor shall maintain accurate records of all deviations in work as actually installed from work indicated on the drawings. On completion of the project, two (2) complete sets of marked-up prints shall be delivered to the Architect.

11. Inspection and Certificates
    a. On the completion of the entire installation, the approval of the Architect and Owner shall be secured, covering the installation throughout. The Contractor shall obtain and pay for Certificate of Approval from the public authorities having jurisdiction. A final inspection certificate shall be submitted to the Architect prior to final payment. Any and all cost incurred for fees shall be paid for by the Contractor.

12. Tests
    a. The right is reserved to inspect and test any portion of the equipment and/or materials during the progress of its erection. This Contractor shall test all wiring and connections for continuity and grounds, before connecting any fixtures or equipment.
    b. The Contractor shall test the entire system in the presence of the Architect or his engineer when the work is finally completed to ensure that all portions are free from short circuits or grounds. All equipment necessary to conduct these tests shall be furnished at the Contractor's expense.

13. Equivalents
    a. When material or equipment is mentioned by name, it shall form the basis of the Contract. When approved by the Architect in writing, other material and equipment may be used in place of those specified, but written application for such substitutions shall be made to the Architect as described in the Bidding Documents. The difference in cost of substitute material or equipment shall be given when such request is made. Approval of substitute is, of course, contingent on same meeting specified requirements and being of such design and dimensions as to comply with space requirements.

14. Guarantee
    a. The Electrical Contractor shall guarantee, by his acceptance of the Contract, that all work installed will be free from defects in workmanship and materials. If during the period of one year, or as otherwise specified, from date of Certificate of Completion and acceptance of work, any such defects in workmanship, materials or performance appear, the Contractor shall, without cost to the Owner, remedy such defects within a reasonable time to be specified in notice from Architect. In default, the Owner may have such work done and charge cost to Contractor.

Section 16B—Basic Materials and Workmanship

1. Portions of the sections of the Documents designated by the letters "A," "B," and "C" and "DIVISION ONE—GENERAL REQUIREMENTS" apply to this Division.
2. Consult Index to be certain that set of Documents and Specifications is complete. Report omissions or discrepancies to the Architect.
3. Conduit Material and Workmanship
    a. GENERAL: The Electrical Contractor shall install a complete raceway system as shown in the drawings and stated in other sections of the specifications. All material used in the raceway system shall be new and the proper material for the job. Conduit, couplings and connectors shall be a product of a

reputable manufacturer equal to conduit as manufactured by Triangle Conduit and Cable or National Electric.

b. CONDUIT INSTALLATION

(1) Conduit shall be of ample size to permit the ready insertion and withdrawal of conductors without abrasion. All joints shall be cut square, reamed smooth and drawn up tight.

(2) Concealed conduits shall be run in as direct a manner as possible, and with as long a bend as possible. Exposed conduit shall be run parallel to or at right angles with the lines of the building. All bends shall be made with standard ells, conduit bent to a radius not less than shown in NEC, or screw jointed conduit fittings. All bends shall be free of dents or flattening. Not more than the equivalent of four quarter bends shall be used in any run between terminals and cabinets, or between outlets and junction or pull boxes.

(3) Conduits shall be continuous from outlet to outlet and from outlet to cabinets, junction or pull boxes, and shall enter and be secured at all boxes in such a manner that each system shall be electrically continuous throughout.

(4) A #14 galvanized iron or steel fish wire shall be left in all conduits in which the permanent wiring is not installed.

(5) Where conduits cross building joints, furnish and install O.Z. Electric Manufacturing Company expansion fittings for contraction, expansion and settlement.

(6) Open ends shall be capped with approved manufactured conduit seals as soon as installed and kept capped until conductors are ready to be installed.

(7) Conduit shall be securely fastened to all sheet metal outlets, junction and pull boxes with galvanized locknuts and bushings, care being observed to see that the full number of threads project through to permit the bushing to be drawn tight against the end of conduit, after which the locknut shall be made sufficiently tight to draw the bushings into firm electrical contact with the box.

(8) For all flush-mounted panels there shall be provided and installed 1¼" empty conduit up through wall and turned out above ceiling and one 1¼" empty conduit down to space below floor except where slab is on grade.

c. CONDUIT HANGERS AND SUPPORTS

(1) Conduit throughout the project shall be securely and rigidly supported to the building structure in a neat and workmanlike manner and wherever possible, parallel runs of horizontal conduit shall be grouped together on adjustable trapeze hangers. Support spacing shall not be more than eight feet.

(2) Exposed conduit shall be supported by one-hole malleable iron straps, two-hole straps, suitable beam clamps or split ring conduit hanger with support rod.

(3) Single conduit 1¼" and larger run concealed horizontally shall be supported by suitable beam clamps or split ring conduit hangers with

support rod. Multiple runs of conduit shall be grouped together on trapeze hangers where possible. Vertical runs shall be supported by steel riser clamps.

(4) Conduit one inch and smaller run concealed above a ceiling may be supported directly to the building structure with strap hangers or No. 14 ga. galvanized wire provided the support spacing does not exceed four feet.

4. Outlet Boxes

   a. GENERAL

   (1) Before locating the outlet boxes check all of the architectural drawings for type of construction and to make sure that there is no conflict with other equipment. The outlet boxes shall be symmetrically located according to room layout and shall not interfere with other work or equipment. Also note any detail of the outlets shown on the drawings.

   (2) Outlet boxes shall be made of galvanized sheet steel unless otherwise noted or required by the NEC and shall be of the proper code size for the required number of conductors. Outlet boxes shall be a minimum of 4 inches square unless specifically noted on the drawings with the exception that a box containing only two current-carrying conductors may be smaller. The outlet boxes shall be complete with the approved type of connectors and required accessories.

   (3) The outlet boxes shall be complete with raised device covers as required to accept device installed. All outlet boxes must be securely fastened in position with the exposed edge of the raised device cover set flush with the finished surface. Approved factory-made knockout seals shall be installed where knockouts are not intact. Galvanized outlet boxes shall be manufactured by RACO, STEEL CITY, APPLETON or approved equal.

   (4) Outlet boxes for exposed work shall be handy boxes with handy box covers unless otherwise noted.

   (5) Outlet boxes located on the exterior in damp or wet locations or as otherwise noted shall be threaded cast aluminum device boxes such as CROUSE HINDS type "FS" or "FD."

   b. RECEPTACLE OUTLET BOXES: Wall receptacles shall be mounted approximately 18" above the finished floor (AFF) unless otherwise noted. When the receptacle is mounted in a masonry wall the bottom of the outlet box shall be in line with the bottom of a masonry unit. Receptacles for electric water coolers shall be installed behind the coolers in accordance with manufacturer's recommendations. All receptacle outlet boxes shall be equipped with grounding lead which shall be connected to grounding terminal of the device.

   c. SWITCH OUTLET BOXES: Wall switches shall be mounted approximately 54 inches above the finished floor (AFF) unless otherwise noted. When the switch is mounted in a masonry wall the bottom of the outlet box shall be in line with the bottom of a masonry unit. Where more than two switches are located, the switches shall be mounted in a gang outlet box with gang cover. Dimmer switches shall be individually mounted unless otherwise noted. Switches with pilot lights, switches with overload motor protection and other special switches that will not conveniently fit under gang wall plates may be individually mounted.

    d. LIGHTING FIXTURE OUTLET BOXES: The lighting fixture outlet boxes shall be furnished with the necessary accessories to install the fixture. The supports must be such as not to depend on the outlet box supporting the fixture. The supports for the lighting fixture shall be independent of the ceiling system. All ceiling outlet boxes shall be equipped with raised circular cover plates with their edges set flush with surface of finished ceiling.

5. Pull Boxes

    a. Pull boxes shall be installed at all necessary points, whether indicated on the drawings or not, to prevent injury to the insulation or other damage that might result from pulling resistance for other reasons necessary to proper installation. Pull box locations shall be approved by the Architect prior to installation. Minimum dimensions shall be not less than NEC requirements and shall be increased if necessary for practical reasons or where required to fit a job condition.

    b. All pull boxes shall be constructed of galvanized sheet steel, code gauge, except that no less than 12 gauge shall be used for any box.

    c. Where boxes are used in connection with exposed conduit, plain covers attached to the box with a suitable number of countersunk flat head machine screws may be used.

    d. Where so indicated, certain pull boxes shall be provided with barriers. These pull boxes shall have a single cover plate, and the barriers shall be of the same gauge as the pull box.

    e. Each circuit in pull box shall be marked with a tag guide denoting panels to which it connects.

    f. Exposed pull boxes will not be permitted in the public spaces.

6. Wireways or Wire Troughs

    a. Wireways shall be used where indicated on the drawings and for mounting groups of switches and/or starters. Wireways shall be the standard manufactured product of a company regularly producing wireway and shall not be a local shop-assembled unit. Wireway shall be of the hinged cover type, Underwriters' Listed, and of sizes indicated or as required by NEC. Finish shall be medium light gray enamel over rust inhibitor. Wireways shall be of the raintight construction where required. Wireways shall be General Electric Type HS or approved equal.

7. Conductor Material and Workmanship

    a. GENERAL

        (1) The Electric Contractor shall provide and install a complete wiring system as shown on the drawing or specifications herein. All conductors used in the wiring system shall be soft drawn copper wire having a conductivity of not less than 98% of that of pure copper, with 600-volt rating, unless otherwise noted. Wire shall be as manufactured by General Cable, Triangle or approved equal.

        (2) The wire shall be delivered to the site in their original unbroken packages, plainly marked or tagged as follows: (a) Underwriters' Labels, (b) Size, kind and insulation of the wire, (c) Name of manufacturing company and the trade name of the wire.

b. CONDUCTOR WORKMANSHIP

    (1) Install conductors in all raceways as required, unless otherwise noted, in a neat and workmanlike manner. Telephone conduits and empty conduits, as noted, shall have a No. 14 ga. galvanized pull wire left in place for future use.

    (2) Conductors shall be color-coded in accordance with the National Electric Code. Mains, feeders, and subfeeders shall be tagged in all pull, junction and outlet boxes and in the gutter of panels with approved code type wire markers.

    (3) No lubricant other than powdered soapstone or approved pulling compound may be used to pull conductors.

    (4) At least eight (8) inches of slack wire shall be left in every outlet box whether it be in use or left for future use.

    (5) All conductors and connections shall test free of grounds, shorts and opens before the job is turned over to the Owner.

8. Lugs, Taps and Splices

    a. Joints on branch circuits shall occur only where such circuits divide and shall consist of one through circuit to which shall be spliced the branch from the circuit. In no case shall joints in branch circuits be left for the fixture hanger to make. No splices shall be made in conductor except at outlet boxes, junction boxes, or splice boxes.

    (1) All joints or splices for No. 10 AWG or smaller shall be made with UL approved wire nuts or compression-type connectors.

    (2) All joints or splices for No. 8 AWG or larger shall be made with a mechanical compression connector. After the conductors have been made mechanically and electrically secure, the entire joint or splice shall be covered with Scotch No. 33 tape or approved equal to make the insulation of the joint or splice equal to the insulation of the conductors. The connector shall be UL approved.

9. Access Doors

    a. The Electrical Contractor shall furnish a ladder to the access doors as shown on the drawings or required for access to junction boxes, etc. The doors shall be 12" square, unless otherwise noted, hinged metal door with metal frames. Door and frame shall be not lighter than 16 gauge sheet steel. The access door shall be of the flush type with screwdriver latching device. The frame shall be constructed so that it can be secured to building material as required. The access doors shall be Milcor or equal. Access door and location shall meet the approval of the Architect.

10. Fuses

    a. Fuses manufactured by Buss or Shawmut shall be furnished and installed as required. Motor protection fuses shall be dual element.

11. Cutting and Patching

    a. On new work the Electrical Contractor shall furnish sketches to the General Contractor showing the locations and sizes of all openings and chases, and furnish and locate all sleeves and inserts required for the installation of the electrical work before the walls, floors and roof are built. The Electrical Contractor shall be responsible for the cost of cutting and patching where any electrical items were not installed or where incorrectly sized or located.

The Contractor shall do all drilling required for the installation of his hangers.

b. On alterations and additions to existing projects, the Electrical Contractor shall be responsible for the cost of all cutting and patching, unless otherwise noted.

c. No structural members shall be cut without the approval of the Architect, and all such cutting shall be done in a manner directed by him. All patching shall be performed in a neat and workmanlike manner acceptable to the Architect.

12. Excavation and Backfilling
    a. The Electrical Contractor shall be responsible for excavation, backfill, tamping, shoring, bracing, pumping, street cuts, repairing of finished surface and all protection for safety of persons and property as required for installing a complete electrical system. All excavation and backfill shall conform to the Architectural Section of the specifications.
    b. Excavation shall be made in a manner to provide a uniform bearing for conduit. Where rock is encountered, excavate 3" below conduit grade and fill with gravel to grade.
    c. After required test and inspections, backfill the ditch and tamp. The first foot above the conduit shall be hand-backfilled with rock-free clean earth. The backfill in the ditches on the exterior and interior of the building shall be tamped to 90%. The Electrical Contractor shall be responsible for any ditches that go down.

13. Equipment and Installation Workmanship
    a. All equipment and material shall be new and shall bear the manufacturer's name and trade name. The equipment and material shall be essentially the standard product of a manufacturer regularly engaged in the production of the required type of equipment and shall be the manufacturer's latest approved design.
    b. The Electrical Contractor shall receive and properly store the equipment and material pertaining to the electrical work. The equipment shall be tightly covered and protected against dirt, water, chemical or mechanical injury and theft. The manufacturer's directions shall be followed completely in the delivery, storage, protection and installation of all equipment and materials.
    c. The Electrical Contractor shall provide and install all items necessary for the complete installation of the equipment as recommended or as required by the manufacturer of the equipment or required by code without additional cost to the owner, regardless whether the items are shown on the plans or covered in the Specifications.
    d. It shall be the responsibility of the Electrical Contractor to clean the electrical equipment, make necessary adjustments and place the equipment into operation before turning equipment over to Owner. Any paint that was scratched during construction shall be "touched up" with factory color paint to the satisfaction of the Architect. Any items that were damaged during construction shall be replaced.

14. Concrete Pads, Support and Encasement
    a. The Electrical Contractor shall be responsible for all concrete pads, supports, piers, bases, foundations and encasement required for the electrical equipment and conduit. The concrete pads for the electrical equipment shall be

six (6) inches larger all around than the base of the equipment and minimum of 4 inches thick unless specifically indicated otherwise.

15. Waterproofing
   a. The Electrical Contractor shall provide all flashing, caulking and sleeves required where his items pass through the outside walls or roof. The waterproofing of the openings shall be made absolutely watertight. The method of installation shall conform to the requirements of Division 7—Moisture Control and/or meet the approval of the Architect.

Section 16C—Service Entrance System

1. Portions of the sections of the Documents designated by the letters "A," "B," "C" and "DIVISION ONE—GENERAL REQUIREMENTS" apply to this Division.
2. Consult Index to be certain that set of Documents and Specifications is complete. Report omissions or discrepancies to the Architect.
3. Service Entrance
   a. GENERAL: The Electrical Contractor shall provide and install a complete service entrance system as shown on the drawings or as required for a complete system. All material and workmanship shall conform with Section 16B of the specifications, National Electric Code and the local electric code. The electric service entrance shall conform to the requirements and regulations of the electric utility serving the project.
   b. ELECTRIC UTILITY CHARGE: The Electrical Contractor shall make all arrangements with the electric utility and pay all charges made by the electric utility for permanent electric service to the project. In the event that the electric utility's charges are not available at the time the project is bid, the Electrical Contractor shall qualify his bid to notify the Owner that such charges are not included.
   c. METERING: The Electrical Contractor shall provide and install raceway, install current transformer cabinet and/or meter trim for metering facilities, as required by the electric utility serving the project. The electric utility will provide the meter installation including meter, current transformers and connections.
   d. GROUNDING: The Electrical Contractor shall properly ground the electrical system as required by the National Electric Code.
   e. CONDUIT: The conduit used for service entrance shall be galvanized rigid steel conduit unless otherwise noted on drawings.
   f. CONDUCTORS: Conductors for the service entrance shall be copper type RHW or THW rated at $75°C$ unless otherwise noted. The conductors indicated on the drawings are based on aluminum.

Section 16D—Electrical Distribution System

1. Portions of the sections of the Documents designated by the letters "A," "B," and "C" and "DIVISION ONE—GENERAL REQUIREMENTS" apply to this Division.
2. Consult Index to be certain that set of Documents and Specifications is complete. Report omissions of discrepancies to the Architect.

3. Feeders and Branch Circuits
   a. GENERAL: The Electrical Contractor shall provide and install a complete electrical distribution system as shown on the drawings or as required for a complete system. All materials and workmanship shall conform with Section 16B of the Specifications, National Electrical Code and the local electric code.
   b. CONDUIT MATERIALS
      (1) Rigid Conduit (Heavy Wall): Rigid conduit shall be galvanized rigid steel conduit with a minimum size of ¾ inch unless otherwise stated. Rigid steel conduit shall be installed for the following services and locations: service entrance, underground in contact with earth, in concrete slab, panel feeders, exterior of building walls, motor feeders over 10 hp, electrical equipment feeders over 10 kW, "wet" locations, and as required by the National Electric Code and local codes.
      (2) Electrical Metallic Tubing (EMT): Electrical metallic tubing shall be galvanized steel with a minimum size of ¾ inch. Electrical metallic tubing shall be used in all locations not otherwise specified for rigid or flexible conduit and where not in violation of the National Electric Code.
      (3) Flexible Metal Conduit: Flexible metal conduit shall be galvanized steel. Flexible metal conduit located in wet locations shall be the liquid-tight type. Flexible metal conduit may be used in place of EMT where completely accessible, such as above removable acoustical tile ceilings and for exposed work in unfinished spaces.

      A short piece of flexible metal conduit shall be used for the connection to all motors and vibrating equipment, connection between recessed light fixtures and junction box, and as otherwise noted, provided the use meets the requirements of the National Electric Code and local codes. The flexible metal conduit shall be the type approved for continuous grounding.
   c. CONDUCTOR MATERIAL
      (1) The conductor material shall be as follows, unless otherwise noted:
         (a) Feeders: Shall be type RHW or THW rated at 75°C.
         (b) Branch Circuits: Shall be Type THW rated at 75°C, except branch circuits with conductor sizes of No. 10 and smaller in dry locations may be type TW rated at 60°C.
         (c) Special Locations: Conductors in special locations such as range hoods, lighting fixtures, etc., shall be as required by the National Electrical Code, local code or as otherwise stated.
      (2) No conductor shall be smaller than No. 12 wire, except for the control wiring and as stated in other sections of the Specifications or on the drawings. Wiring to switches shall not be considered as control wiring.
      (3) Conductors indicated on the drawings are based on copper. Panel, motor and electrical equipment feeders with a size of No. 1/0 and larger may be aluminum, providing the size of the conductor is increased to have the same or more current carrying capacity as the copper conductors. Also, the conduit sizes shall be increased accordingly.
      (4) All conductors with the size of No. 8 or larger shall be stranded.

(5) All lighting and receptacle branch circuits in excess of 100 linear feet shall be increased one size to prevent excessive voltage drop.

4. Safety Switches (FSS) (NFSS)

    a. GENERAL: Furnish and install safety switches as indicated on the drawings or as required. All safety switches shall be NEMA General Duty Type and Underwriters' Laboratories Listed. The switches shall be Fused Safety Switches (FSS) or Nonfused Safety Switches (NFSS) as shown on the drawings or required.

    b. SWITCHES: Switches shall have a quick-make and quick-break operating handle and mechanism which shall be an integral part of the box. Padlocking provisions shall be provided for padlocking in the "OFF" position with at least three padlocks. Switches shall be at least horsepower rated for 250 volts AC or DC as required. Lugs shall be UL listed for copper and aluminum cable.

    c. ENCLOSURES: Switches shall be furnished in NEMA 1 general-purpose enclosures with knockouts unless otherwise noted or required. Switches located on the exterior of the building or in "wet" locations shall have NEMA 3R enclosures (WP).

    d. INSTALLATION: The safety switches shall be securely mounted in accordance with the NEC. The Contractor shall provide all mounting materials. Install fuses in the FSS. The fuses shall be dual element on motor circuits.

    e. MANUFACTURER: Square D, General Electric, Cutler-Hammer, Westinghouse, or ITE.

5. Panel Boards—Circuit Breaker

    a. GENERAL: Furnish and install circuit-breaker panel boards as indicated in the panelboard schedule and where shown on the drawings. The panel board shall be dead front safety type equipped with molded case circuit breakers and shall be the type as listed in the panelboard schedule: Service entrance panel boards shall include a full capacity box bonding strap and approved for service entrance. The acceptable manufacturers of the panel boards are ITE, General Electric, Cutler-Hammer, and Westinghouse, provided that they are fully equal to the type listed on the drawings. The panel board shall be listed by Underwriters' Laboratories and bear the UL Label.

    b. CIRCUIT BREAKERS: Provide molded case circuit breakers of frame, trip rating and interrupting capacity as shown on the schedule. Also, provide the number of spaces for future circuit breakers as shown in the schedule. The circuit breakers shall be quick-make, quick-break, thermal-magnetic, trip indicating and have common trip on all multipole breakers with internal tie mechanism.

    c. PANELBOARD BUS ASSEMBLY: Bus bar connections to the branch circuit breakers shall be the "phase sequence" type. Single-phase 3-wire panelboard busing shall be such that any two adjacent single-pole breakers are connected to opposite polarities in such a manner that 2-pole breakers can be installed in any location. Three-phase 4-wire breakers are individually connected to each of the three different phases in such a manner that 2- or 3-pole breakers can be installed at any location. All current carrying parts of the bus assembly shall be plated. Mains ratings shall be as shown in the panelboard schedule on the plans. Provide solid neutral assembly (S/N) when required.

    d. WIRING TERMINALS: Terminals for feeder conductors to the panelboard mains and neutral shall be suitable for the type of conductor specified. Terminals for branch circuit wiring, both breaker and neutral, shall be suitable for the type of conductor specified.

    e. CABINETS AND FRONTS: The panelboard bus assembly shall be enclosed in a steel cabinet. The size of the wiring gutters and gauge of steel shall be in accordance with NEMA Standards. The box shall be fabricated from galvanized steel or equivalent rust resistant steel. Fronts shall include door and have flush, brushed stainless steel, spring-loaded door pulls. The flush lock shall not protrude beyond the front of the door. All panelboard locks shall be keyed alike. Fronts shall not be removable with door in the locked position.

    f. DIRECTORY: On the inside of the door of each cabinet, provide a typewritten directory which will indicate the location of the equipment or outlets supplied by each circuit. The directory shall be mounted in a metal frame with a nonbreakable transparent cover. The panelboard designation shall be typed on the directory card and panel designation stenciled in 1½" high letters on the inside of the door.

    g. PANEL BOARD INSTALLATION

       (1) Before installing panel boards check all of the architectural drawings for possible conflict of space and adjust the location of the panel board to prevent such conflict with other items.

       (2) When the panel board is recessed into a wall serving an area with accessible ceiling space, provide and install an empty conduit system for future wiring. All 1¼" space above the panel board and under the panel board if such accessible ceiling space exists.

       (3) The panel boards shall be mounted in accordance with Article 373 of the NEC. The Electrical Contractor shall furnish all material for mounting the panel boards.

6. Wiring Devices

    a. GENERAL: The wiring devices specified below with Arrow Hart numbers may also be the equivalent wiring device as manufactured by Bryant Electric, Harvey Hubbell, or Pass & Seymour. All other items shall be as specified.

    b. WALL SWITCHES: Where more than one flush wall switch is indicated in the same location, the switches shall be mounted in gangs under a common plate.

(1)	Single-Pole	AH#1991
(2)	Three-way	AH#1993
(3)	Four-way	AH#1994
(4)	Switch with pilot light	AH#2999-R
(5)	Motor switch—Surface	AH#6808
(6)	Motor switch—Flush	AH#6808-F

    c. RECEPTACLES

(1)	Duplex	AH#6739
(2)	Clock outlet	AH#5708
(3)	Weatherproof	AH#5735-F

       (4) Floor receptacles—Steel City Series 600 floor box with bronze edge ring, floor plate P-60-1, bronze carpet plate and service fitting SFH-40-RG.

(5) Floor outlet for Telephone and Alarm—Steel City Series 600 floor box with bronze edge ring, floor plate P-60-1, bronze carpet plate and service fitting SFL-10.

   d. WALL PLATE: Stainless steel wall plates with satin finish minimum 0.030 inch shall be provided for all outlets and switches.

Section 16E—Lighting Fixtures and Lamps

1. Portions of the sections of the Documents designated by the letters "A," "B," and "C" and "DIVISION ONE—GENERAL REQUIREMENTS" apply to this division.
2. Consult Index to be certain that set of Documents and Specifications is complete. Report omissions or discrepancies to the Architect.
3. Lighting Fixtures
   a. GENERAL: The Electrical Contractor shall furnish, install and connect all lighting fixtures to the building wiring system unless otherwise noted.
   b. FIXTURE TYPE: The fixture for each location is indicated by type letter. Refer to fixture schedule on the drawings for each type, manufacturer, catalog number and type of mounting.
   c. FLUORESCENT BALLASTS: All fluorescent fixtures shall have ETL-CBM high power factor, quiet operating, Class "A" sound rated, thermally protected Class "P" cool-rated ballast with UL approval. Ballasts shall be as manufactured by General Electric, Advance Jefferson or approved equal. The ballasts shall be subject to a two (2) year manufacturer's guarantee. Guarantee shall be filed with the Owner.
   d. SHOP DRAWINGS
       (1) Shop drawings for lighting fixtures shall indicate each type together with manufacturer's name and catalog number, complete photometric data compiled by an independent testing laboratory and type of lamp(s) to be installed. No fixtures shall be delivered to the job until approved by the Architect.
       (2) If the Electrical Contractor submits shop drawings on a fixture for approval other than those specified, he shall also submit a sample fixture when requested by the Architect. The sample fixture will be returned to the Electrical Contractor. The decision of the Architect shall be final.
   e. COORDINATION: It shall be the responsibility of the Electrical Contractor to coordinate with the ceiling contractor and the General Contractor in order that the proper type fixture be furnished to match the ceiling suspension system being installed or building construction material.
4. Lamps
   a. The Electrical Contractor shall furnish and install lamps in all fixtures as indicated on the drawings or as required. Fluorescent lamps shall be standard cool white and incandescent lamps shall be inside frosted unless otherwise noted on the drawings.
   b. Lamps shall be manufactured by General Electric, Westinghouse or Sylvania.

Section 16F—Special Systems

1. Portions of the sections of the Documents designated by the letters "A," "B," and "C" and "DIVISION ONE—GENERAL REQUIREMENTS" apply to this division.
2. Consult Index to be certain that set of Documents and Specifications is complete. Report omissions or discrepancies to the Architect.
3. Telephone Raceway Systems
   a. GENERAL: The Electrical Contractor shall provide and install empty raceway, outlet boxes, pull boxes and associated equipment required for a complete telephone system as indicated on the drawings and specified herein. All materials and workmanship shall conform with Section 16B of the Specifications. All wiring shall be installed by the local telephone company. The entire installation shall be in accordance with the requirements of the local telephone company.
   b. RIGID CONDUIT (Heavy Wall): Rigid conduit shall be installed in the following locations: service entrance, underground in contact with earth, in concrete slab and "wet" locations.
   c. ELECTRIC METALLIC CONDUIT (EMT): Electric metallic tubing shall be used in all locations not otherwise specified to be rigid conduit.
   d. OUTLETS: Telephone wall outlets shall consist of a 4" two-gang outlet box, raised device cover and telephone device plate of the same material as the receptacle device plates. The conduit shall extend from the outlet to the designated telephone space unless otherwise noted.
   e. PULL WIRE: The Electrical Contractor shall install a No. 14 ga. galvanized pull wire in the raceway system for future use.
   f. MOUNTING HEIGHTS: The wall outlets shall be mounted at approximately the following heights unless otherwise noted on the drawings or required by telephone company: Desk Phones—18" AFF, Wall Phone—58" AFF, Telephone Booth—7'6" AFF.
4. Emergency Lighting System
   a. The Electrical Contractor shall provide and install a complete emergency lighting system as indicated on the drawings and specified herein. The system shall originate on the line side of the service entrance main switch, through overcurrent protective equipment to each exit light fixture and each fixture designated as being "emergency light." The switch shall be painted red. The Contractor shall be responsible for verification with local governing authorities of the proper letter and background colors of exit light fixtures before purchase of same. The entire installation shall be in accordance with the National Electric Code, the local electric code and the fire protection department having authority in the local jurisdiction.

Appendix A
Coefficients of Utilization

PRINCIPLES OF ILLUMINATION

COEFFICIENTS OF UTILIZATION

Reflectances — Ceiling cavity 80%, 50%, 10%, 0%; Walls 50%, 30%, 10% (and 0%). Values are Coefficients of utilization.

Category III — Ventilated dome reflector — Distribution 0 / 70 — Spacing not to exceed: 1.3 × mounting height

RCR	80% 50	80% 30	80% 10	50% 50	50% 30	50% 10	10% 50	10% 30	10% 10	0%
1	0.85	0.82	0.79	0.79	0.77	0.75	0.73	0.72	0.71	0.69
2	0.74	0.69	0.65	0.70	0.66	0.62	0.65	0.62	0.59	0.58
3	0.65	0.60	0.54	0.62	0.57	0.53	0.57	0.54	0.51	0.49
4	0.58	0.51	0.46	0.55	0.49	0.45	0.51	0.47	0.44	0.42
5	0.50	0.44	0.38	0.47	0.42	0.37	0.45	0.40	0.36	0.35
6	0.44	0.38	0.33	0.43	0.36	0.32	0.40	0.35	0.32	0.30
7	0.40	0.33	0.28	0.38	0.33	0.28	0.36	0.32	0.27	0.26
8	0.36	0.29	0.24	0.34	0.28	0.24	0.32	0.27	0.23	0.22
9	0.33	0.26	0.20	0.31	0.25	0.20	0.29	0.24	0.20	0.18
10	0.29	0.22	0.18	0.28	0.22	0.18	0.26	0.21	0.18	0.17

Category I — R-52 Filament reflector lamp wide dist. — 500- and 750-watt — Distribution 0 / 100 — Spacing not to exceed: 1.5 × mounting height

RCR	80% 50	80% 30	80% 10	50% 50	50% 30	50% 10	10% 50	10% 30	10% 10	0%
1	1.08	1.05	1.02	1.01	0.99	0.97	0.94	0.93	0.91	0.89
2	0.98	0.93	0.89	0.93	0.89	0.86	0.88	0.85	0.82	0.80
3	0.89	0.83	0.78	0.85	0.80	0.76	0.80	0.76	0.73	0.71
4	0.81	0.74	0.68	0.77	0.72	0.67	0.73	0.69	0.65	0.64
5	0.73	0.66	0.60	0.70	0.64	0.59	0.66	0.62	0.58	0.56
6	0.67	0.59	0.53	0.64	0.58	0.52	0.61	0.56	0.52	0.50
7	0.60	0.52	0.47	0.58	0.51	0.46	0.55	0.50	0.46	0.45
8	0.54	0.46	0.40	0.52	0.45	0.40	0.49	0.44	0.40	0.38
9	0.48	0.40	0.35	0.46	0.39	0.35	0.44	0.38	0.34	0.33
10	0.43	0.36	0.30	0.42	0.35	0.30	0.40	0.34	0.30	0.28

Category I — R-67 Filament reflector lamp narrow dist. — 500- and 750-watt — Distribution 0 / 100 — Spacing not to exceed: 0.6 × mounting height

RCR	80% 50	80% 30	80% 10	50% 50	50% 30	50% 10	10% 50	10% 30	10% 10	0%
1	1.10	1.08	1.05	1.04	1.02	1.00	0.97	0.96	0.95	0.93
2	1.02	0.98	0.94	0.97	0.94	0.91	0.91	0.89	0.88	0.86
3	0.95	0.90	0.85	0.91	0.87	0.83	0.86	0.83	0.81	0.79
4	0.88	0.82	0.78	0.85	0.80	0.76	0.81	0.77	0.75	0.73
5	0.82	0.76	0.71	0.79	0.74	0.70	0.76	0.72	0.69	0.67
6	0.77	0.70	0.66	0.74	0.69	0.65	0.72	0.68	0.64	0.63
7	0.71	0.65	0.61	0.69	0.64	0.60	0.67	0.63	0.60	0.58
8	0.66	0.60	0.56	0.65	0.59	0.55	0.63	0.58	0.55	0.54
9	0.62	0.55	0.51	0.60	0.55	0.51	0.60	0.54	0.50	0.49
10	0.58	0.51	0.47	0.56	0.51	0.47	0.55	0.50	0.46	0.45

Category III — Ventilated porcelain enamel low bay, 400-W phos. coated vapor lamp — Distribution 0 / 76 — Spacing not to exceed: 1.2 × mounting height

RCR	80% 50	80% 30	80% 10	50% 50	50% 30	50% 10	10% 50	10% 30	10% 10	0%
1	0.81	0.78	0.76	0.76	0.74	0.72	0.71	0.69	0.68	0.67
2	0.73	0.69	0.65	0.69	0.66	0.63	0.64	0.62	0.60	0.59
3	0.65	0.60	0.56	0.62	0.58	0.55	0.58	0.55	0.53	0.51
4	0.59	0.53	0.49	0.56	0.52	0.48	0.53	0.50	0.47	0.45
5	0.53	0.47	0.43	0.51	0.46	0.42	0.48	0.44	0.41	0.40
6	0.48	0.42	0.38	0.46	0.41	0.37	0.44	0.40	0.37	0.35
7	0.43	0.37	0.33	0.41	0.36	0.32	0.39	0.36	0.32	0.31
8	0.39	0.33	0.29	0.38	0.32	0.28	0.36	0.32	0.28	0.28
9	0.36	0.30	0.26	0.34	0.29	0.25	0.33	0.28	0.25	0.24
10	0.32	0.27	0.23	0.31	0.26	0.23	0.30	0.25	0.22	0.21

Category III — 18" Ventilated alum. high bay conc. dist. 400-W clear vapor lamp — Distribution 9 / 77 — Spacing not to exceed: 0.7 × mounting height

RCR	80% 50	80% 30	80% 10	50% 50	50% 30	50% 10	10% 50	10% 30	10% 10	0%
1	0.93	0.90	0.88	0.85	0.83	0.82	0.76	0.75	0.74	0.72
2	0.86	0.82	0.79	0.79	0.77	0.74	0.72	0.70	0.69	0.67
3	0.79	0.75	0.71	0.74	0.70	0.68	0.68	0.65	0.64	0.62
4	0.74	0.69	0.65	0.69	0.65	0.62	0.64	0.61	0.59	0.57
5	0.68	0.63	0.59	0.64	0.60	0.57	0.60	0.57	0.54	0.53
6	0.63	0.58	0.54	0.60	0.56	0.52	0.56	0.53	0.50	0.49
7	0.59	0.53	0.49	0.56	0.51	0.48	0.52	0.49	0.46	0.45
8	0.55	0.49	0.45	0.52	0.47	0.44	0.49	0.45	0.43	0.41
9	0.50	0.45	0.41	0.48	0.43	0.40	0.45	0.42	0.39	0.38
10	0.47	0.41	0.38	0.45	0.40	0.37	0.42	0.38	0.36	0.35

Category III — 18" Ventilated alum. high bay spread dist. 400-W coated vapor lamp — Distribution 10 / 74 — Spacing not to exceed: 1.2 × mounting height

RCR	80% 50	80% 30	80% 10	50% 50	50% 30	50% 10	10% 50	10% 30	10% 10	0%
1	0.88	0.86	0.84	0.80	0.79	0.77	0.71	0.70	0.69	0.67
2	0.81	0.77	0.74	0.75	0.72	0.70	0.67	0.65	0.64	0.62
3	0.74	0.70	0.66	0.69	0.65	0.62	0.62	0.60	0.58	0.56
4	0.68	0.63	0.59	0.64	0.60	0.57	0.58	0.55	0.53	0.51
5	0.63	0.57	0.53	0.59	0.55	0.51	0.54	0.51	0.49	0.47
6	0.58	0.52	0.48	0.54	0.50	0.46	0.50	0.47	0.44	0.43
7	0.53	0.47	0.43	0.50	0.45	0.42	0.46	0.43	0.40	0.39
8	0.48	0.43	0.39	0.46	0.41	0.38	0.42	0.39	0.36	0.35
9	0.44	0.39	0.35	0.42	0.37	0.34	0.39	0.35	0.33	0.31
10	0.41	0.35	0.31	0.39	0.34	0.30	0.36	0.32	0.30	0.28

Category III — 24" Ventilated porcelain enamel 1000-W phosphor coated vapor lamp — Distribution 11 / 73 — Spacing not to exceed: 1.3 × mounting height

RCR	80% 50	80% 30	80% 10	50% 50	50% 30	50% 10	10% 50	10% 30	10% 10	0%
1	0.86	0.83	0.80	0.78	0.76	0.73	0.68	0.67	0.65	0.63
2	0.77	0.72	0.68	0.70	0.66	0.63	0.61	0.58	0.57	0.55
3	0.68	0.62	0.57	0.62	0.58	0.54	0.55	0.52	0.49	0.47
4	0.61	0.55	0.49	0.56	0.51	0.47	0.50	0.46	0.43	0.41
5	0.55	0.48	0.42	0.50	0.45	0.41	0.45	0.41	0.38	0.36
6	0.49	0.42	0.37	0.45	0.39	0.35	0.40	0.36	0.33	0.31
7	0.43	0.36	0.31	0.40	0.34	0.30	0.36	0.31	0.28	0.26
8	0.39	0.32	0.28	0.36	0.30	0.26	0.32	0.28	0.25	0.23
9	0.35	0.28	0.24	0.33	0.27	0.23	0.29	0.25	0.22	0.20
10	0.32	0.25	0.21	0.29	0.24	0.20	0.26	0.22	0.19	0.17

COEFFICIENTS OF UTILIZATION

Luminaire	Distribution	Spacing not to exceed	Ceiling cavity / RCR	Reflectances 80%			50%			10%			0%
			Walls	50%	30%	10%	50%	30%	10%	50%	30%	10%	0%
Category III — 24" Ventilated alum. high bay spread dist. 1000-W phos. ctd. vapor lamp	7 / 79	1.0 × mounting height	1	0.91	0.88	0.86	0.84	0.82	0.80	0.75	0.74	0.73	0.71
			2	0.83	0.78	0.75	0.77	0.73	0.71	0.70	0.67	0.66	0.64
			3	0.75	0.69	0.65	0.70	0.65	0.62	0.64	0.61	0.58	0.56
			4	0.68	0.62	0.57	0.63	0.58	0.55	0.58	0.55	0.52	0.50
			5	0.61	0.55	0.50	0.57	0.52	0.48	0.53	0.49	0.46	0.44
			6	0.55	0.49	0.44	0.52	0.47	0.43	0.48	0.44	0.41	0.39
			7	0.50	0.43	0.38	0.47	0.41	0.37	0.43	0.39	0.36	0.34
			8	0.45	0.39	0.34	0.43	0.37	0.33	0.39	0.35	0.32	0.30
			9	0.41	0.34	0.30	0.39	0.33	0.29	0.36	0.32	0.28	0.27
			10	0.37	0.31	0.27	0.35	0.30	0.26	0.33	0.28	0.25	0.24
Category III — 24" Ventilated alum. high bay 1000-W phos. coated vapor lamp	12 / 73	1.3 × mounting height	1	0.90	0.88	0.86	0.81	0.80	0.78	0.71	0.70	0.70	0.67
			2	0.83	0.79	0.76	0.76	0.73	0.71	0.67	0.66	0.64	0.62
			3	0.77	0.72	0.68	0.70	0.67	0.64	0.63	0.61	0.59	0.57
			4	0.71	0.66	0.62	0.66	0.62	0.59	0.59	0.57	0.55	0.53
			5	0.65	0.60	0.56	0.61	0.57	0.53	0.55	0.52	0.50	0.48
			6	0.60	0.55	0.50	0.56	0.52	0.48	0.52	0.48	0.46	0.44
			7	0.55	0.50	0.46	0.52	0.47	0.44	0.48	0.44	0.42	0.40
			8	0.51	0.45	0.41	0.48	0.43	0.40	0.44	0.41	0.38	0.37
			9	0.47	0.41	0.38	0.44	0.40	0.37	0.41	0.38	0.35	0.34
			10	0.44	0.38	0.34	0.41	0.37	0.33	0.38	0.35	0.32	0.31
Category III — 2 T-12 Lamps — any loading for T-10 lamps — C.U. × 1.02	10 / 75	1.3 × mounting height	1	0.88	0.84	0.81	0.79	0.77	0.74	0.69	0.68	0.66	0.64
			2	0.77	0.71	0.66	0.70	0.65	0.62	0.61	0.59	0.56	0.54
			3	0.68	0.61	0.56	0.61	0.56	0.52	0.54	0.51	0.48	0.46
			4	0.60	0.52	0.47	0.54	0.49	0.44	0.48	0.44	0.41	0.39
			5	0.52	0.45	0.39	0.48	0.42	0.37	0.43	0.38	0.35	0.33
			6	0.47	0.39	0.34	0.43	0.37	0.32	0.38	0.34	0.30	0.28
			7	0.42	0.34	0.29	0.38	0.32	0.28	0.34	0.30	0.26	0.24
			8	0.37	0.30	0.25	0.34	0.28	0.24	0.31	0.26	0.22	0.21
			9	0.33	0.26	0.21	0.31	0.25	0.21	0.28	0.23	0.19	0.18
			10	0.30	0.23	0.19	0.28	0.22	0.18	0.25	0.20	0.17	0.15
Category II — 2 T-12 Lamps — any loading center shield for T-10 lamps — C.U. × 1.02	17 / 71	1.3 × mounting height	1	0.88	0.85	0.81	0.77	0.75	0.73	0.65	0.64	0.62	0.59
			2	0.77	0.71	0.67	0.68	0.64	0.60	0.57	0.55	0.53	0.50
			3	0.68	0.61	0.56	0.60	0.55	0.51	0.51	0.48	0.45	0.42
			4	0.60	0.53	0.47	0.53	0.48	0.43	0.45	0.42	0.38	0.36
			5	0.53	0.45	0.40	0.47	0.41	0.36	0.40	0.36	0.33	0.30
			6	0.47	0.39	0.34	0.42	0.36	0.31	0.36	0.31	0.28	0.26
			7	0.42	0.34	0.29	0.38	0.31	0.27	0.32	0.28	0.24	0.22
			8	0.38	0.30	0.25	0.34	0.28	0.23	0.29	0.24	0.21	0.19
			9	0.34	0.26	0.22	0.30	0.24	0.20	0.26	0.21	0.18	0.16
			10	0.31	0.24	0.19	0.26	0.22	0.18	0.24	0.19	0.16	0.14
Category II — 2 T-12 Lamps — any loading center shield for T-10 lamps — C.U. × 1.02	18 / 66	1.3 × mounting height	1	0.84	0.81	0.78	0.74	0.72	0.70	0.61	0.60	0.59	0.56
			2	0.75	0.70	0.65	0.66	0.62	0.59	0.55	0.53	0.51	0.48
			3	0.66	0.60	0.56	0.59	0.54	0.51	0.49	0.47	0.44	0.42
			4	0.59	0.52	0.47	0.52	0.47	0.43	0.44	0.41	0.38	0.36
			5	0.52	0.45	0.40	0.46	0.41	0.37	0.39	0.36	0.33	0.31
			6	0.47	0.40	0.35	0.42	0.36	0.32	0.36	0.32	0.29	0.27
			7	0.42	0.35	0.30	0.37	0.32	0.28	0.32	0.28	0.25	0.23
			8	0.38	0.31	0.26	0.34	0.28	0.24	0.29	0.25	0.22	0.20
			9	0.34	0.27	0.22	0.30	0.25	0.21	0.26	0.22	0.19	0.17
			10	0.31	0.24	0.20	0.27	0.22	0.18	0.23	0.19	0.17	0.15
Category III — 3 T-12 Lamps — 430 or 800 MA for T-10 lamps — C.U. × 1.02	9 / 74	1.3 × mounting height	1	0.86	0.83	0.80	0.78	0.76	0.73	0.69	0.67	0.66	0.64
			2	0.75	0.70	0.66	0.69	0.65	0.61	0.61	0.58	0.56	0.54
			3	0.67	0.60	0.55	0.61	0.56	0.52	0.54	0.51	0.48	0.46
			4	0.59	0.52	0.47	0.54	0.49	0.44	0.48	0.45	0.41	0.39
			5	0.52	0.45	0.39	0.48	0.42	0.38	0.43	0.39	0.35	0.33
			6	0.46	0.39	0.34	0.43	0.37	0.32	0.38	0.34	0.30	0.28
			7	0.41	0.34	0.29	0.38	0.32	0.28	0.34	0.30	0.26	0.25
			8	0.37	0.30	0.25	0.34	0.28	0.24	0.31	0.26	0.23	0.21
			9	0.33	0.26	0.22	0.31	0.25	0.21	0.28	0.23	0.20	0.18
			10	0.30	0.23	0.19	0.28	0.22	0.18	0.25	0.21	0.17	0.16
Category II — 3 T-12 Lamps — 430 or 800 MA for T-10 lamps — C.U. × 1.02	15 / 69	1.3 × mounting height	1	0.85	0.82	0.79	0.76	0.73	0.71	0.64	0.63	0.62	0.59
			2	0.75	0.70	0.65	0.67	0.63	0.59	0.57	0.55	0.52	0.50
			3	0.66	0.60	0.55	0.59	0.54	0.50	0.51	0.48	0.45	0.42
			4	0.59	0.52	0.46	0.52	0.47	0.43	0.45	0.41	0.38	0.36
			5	0.51	0.44	0.39	0.46	0.40	0.36	0.40	0.36	0.33	0.30
			6	0.46	0.39	0.33	0.41	0.35	0.31	0.36	0.31	0.28	0.26
			7	0.41	0.34	0.29	0.37	0.32	0.27	0.32	0.28	0.24	0.23
			8	0.37	0.30	0.25	0.33	0.27	0.23	0.29	0.24	0.21	0.19
			9	0.33	0.26	0.21	0.30	0.24	0.20	0.26	0.21	0.18	0.16
			10	0.30	0.23	0.19	0.27	0.21	0.18	0.23	0.19	0.16	0.14

COEFFICIENTS OF UTILIZATION													
				Reflectances									
Luminaire	Distribution	Spacing not to exceed	Ceiling cavity	80%			50%			10%			0%
			Walls	50%	30%	10%	50%	30%	10%	50%	30%	10%	0%
			RCR	Coefficients of utilization									
Category V — 2 T-12 Lamps − 430 MA for 800 MA − C.U. × 0.96	12 / 60	1.5 × mounting height	1	0.70	0.66	0.63	0.62	0.59	0.57	0.52	0.51	0.49	0.47
			2	0.60	0.54	0.50	0.53	0.49	0.46	0.45	0.42	0.40	0.37
			3	0.52	0.46	0.41	0.46	0.41	0.38	0.39	0.36	0.33	0.31
			4	0.46	0.39	0.34	0.41	0.36	0.32	0.35	0.31	0.28	0.26
			5	0.40	0.33	0.28	0.36	0.30	0.26	0.31	0.27	0.24	0.22
			6	0.36	0.29	0.24	0.32	0.26	0.22	0.27	0.23	0.20	0.18
			7	0.32	0.25	0.21	0.29	0.23	0.19	0.25	0.21	0.17	0.16
			8	0.29	0.22	0.18	0.26	0.20	0.17	0.22	0.18	0.15	0.13
			9	0.26	0.19	0.15	0.23	0.18	0.14	0.20	0.16	0.13	0.11
			10	0.23	0.17	0.13	0.21	0.16	0.12	0.18	0.14	0.11	0.10
Category V — 2 T-12 Lamps − 430 MA prismatic lens 1' wide − for T-10 lamps − C.U. × 1.02	0 / 59	1.2 × mounting height	1	0.63	0.61	0.59	0.59	0.58	0.56	0.55	0.54	0.53	0.52
			2	0.57	0.54	0.51	0.54	0.51	0.49	0.50	0.49	0.47	0.46
			3	0.51	0.48	0.44	0.49	0.46	0.43	0.46	0.44	0.42	0.41
			4	0.46	0.42	0.39	0.44	0.41	0.38	0.42	0.39	0.37	0.36
			5	0.42	0.37	0.34	0.40	0.36	0.34	0.38	0.35	0.33	0.32
			6	0.38	0.34	0.30	0.37	0.33	0.30	0.35	0.32	0.29	0.28
			7	0.35	0.30	0.27	0.33	0.29	0.27	0.32	0.29	0.26	0.25
			8	0.31	0.27	0.24	0.30	0.26	0.23	0.29	0.26	0.23	0.22
			9	0.28	0.24	0.21	0.27	0.23	0.20	0.26	0.23	0.20	0.19
			10	0.26	0.21	0.18	0.25	0.21	0.18	0.24	0.20	0.18	0.17
Category V — 2 T-12 Lamps − 430 MA prismatic lens 2' wide − for T-10 lamps − C.U. × 1.01	0 / 68	1.2 × mounting height	1	0.73	0.71	0.68	0.69	0.67	0.66	0.64	0.62	0.61	0.60
			2	0.66	0.62	0.59	0.62	0.59	0.57	0.58	0.56	0.55	0.53
			3	0.59	0.55	0.51	0.56	0.53	0.50	0.53	0.50	0.48	0.47
			4	0.53	0.48	0.45	0.51	0.47	0.44	0.48	0.45	0.43	0.41
			5	0.48	0.43	0.39	0.46	0.42	0.39	0.44	0.40	0.38	0.36
			6	0.44	0.38	0.34	0.42	0.37	0.34	0.40	0.36	0.33	0.32
			7	0.39	0.34	0.30	0.38	0.33	0.30	0.36	0.32	0.30	0.28
			8	0.36	0.30	0.26	0.34	0.30	0.26	0.33	0.29	0.26	0.25
			9	0.32	0.27	0.23	0.31	0.26	0.23	0.29	0.25	0.23	0.21
			10	0.29	0.24	0.20	0.28	0.23	0.20	0.27	0.23	0.20	0.19
Category V — 4 T-12 Lamps − 430 MA prismatic lens 2' wide − for T-10 lamps − C.U. × 1.02	0 / 62	1.2 × mounting height	1	0.66	0.64	0.62	0.62	0.61	0.59	0.58	0.57	0.56	0.55
			2	0.60	0.56	0.53	0.56	0.54	0.52	0.53	0.51	0.49	0.48
			3	0.54	0.50	0.46	0.51	0.48	0.45	0.48	0.46	0.44	0.43
			4	0.49	0.44	0.41	0.46	0.43	0.40	0.44	0.41	0.39	0.38
			5	0.44	0.39	0.35	0.42	0.38	0.35	0.40	0.37	0.34	0.33
			6	0.40	0.35	0.31	0.38	0.34	0.31	0.36	0.33	0.31	0.29
			7	0.36	0.31	0.28	0.35	0.30	0.27	0.33	0.30	0.27	0.26
			8	0.32	0.28	0.24	0.31	0.27	0.24	0.30	0.26	0.24	0.23
			9	0.29	0.24	0.21	0.28	0.24	0.21	0.27	0.23	0.21	0.20
			10	0.27	0.22	0.19	0.26	0.23	0.19	0.25	0.21	0.18	0.17
Category V — 6 T-12 Lamps − 430 MA prismatic lens 2' wide − for T-10 lamps − C.U. × 1.03	0 / 56	1.2 × mounting height	1	0.60	0.58	0.56	0.56	0.55	0.54	0.52	0.51	0.50	0.49
			2	0.54	0.51	0.48	0.51	0.49	0.47	0.48	0.46	0.45	0.44
			3	0.49	0.45	0.42	0.46	0.43	0.41	0.44	0.41	0.40	0.39
			4	0.44	0.40	0.37	0.42	0.39	0.36	0.40	0.37	0.35	0.34
			5	0.40	0.35	0.32	0.38	0.35	0.32	0.36	0.33	0.31	0.30
			6	0.36	0.32	0.29	0.35	0.31	0.28	0.33	0.30	0.28	0.27
			7	0.33	0.28	0.25	0.32	0.28	0.25	0.30	0.27	0.25	0.24
			8	0.30	0.25	0.22	0.28	0.25	0.22	0.27	0.24	0.22	0.21
			9	0.27	0.22	0.19	0.26	0.22	0.19	0.25	0.21	0.19	0.18
			10	0.24	0.20	0.17	0.23	0.20	0.17	0.22	0.19	0.17	0.16
Category V — 8 T-12 Lamps − 430 MA prismatic lens 4' × 4' − for T-10 lamps − C.U. × 1.02	0 / 55	1.3 × mounting height	1	0.59	0.57	0.55	0.55	0.54	0.52	0.51	0.50	0.49	0.48
			2	0.53	0.50	0.47	0.50	0.48	0.46	0.47	0.45	0.44	0.43
			3	0.48	0.44	0.41	0.45	0.42	0.40	0.43	0.40	0.39	0.38
			4	0.43	0.39	0.36	0.41	0.38	0.35	0.39	0.36	0.34	0.33
			5	0.39	0.35	0.31	0.37	0.34	0.31	0.35	0.32	0.30	0.29
			6	0.35	0.31	0.28	0.34	0.30	0.28	0.32	0.29	0.27	0.26
			7	0.32	0.28	0.25	0.31	0.27	0.25	0.29	0.26	0.24	0.23
			8	0.29	0.25	0.22	0.28	0.24	0.22	0.27	0.24	0.22	0.20
			9	0.26	0.22	0.19	0.25	0.21	0.19	0.24	0.21	0.19	0.18
			10	0.24	0.20	0.17	0.23	0.19	0.17	0.22	0.19	0.17	0.16
Category V — 4 T-12 Lamps − 430 MA prismatic lens 2' wide − for T-10 lamps − C.U. × 1.02	2 / 51	1.2 × mounting height	1	0.56	0.54	0.52	0.52	0.50	0.49	0.47	0.46	0.45	0.44
			2	0.50	0.47	0.45	0.47	0.44	0.42	0.43	0.41	0.40	0.39
			3	0.45	0.41	0.38	0.42	0.39	0.37	0.39	0.37	0.35	0.34
			4	0.41	0.37	0.34	0.38	0.35	0.32	0.35	0.33	0.31	0.30
			5	0.37	0.32	0.29	0.34	0.31	0.28	0.32	0.29	0.27	0.26
			6	0.33	0.29	0.26	0.31	0.28	0.25	0.29	0.27	0.24	0.23
			7	0.30	0.26	0.23	0.29	0.25	0.22	0.27	0.24	0.22	0.20
			8	0.27	0.23	0.20	0.26	0.22	0.20	0.24	0.21	0.19	0.18
			9	0.25	0.20	0.18	0.23	0.20	0.17	0.22	0.19	0.17	0.16
			10	0.22	0.18	0.16	0.21	0.18	0.15	0.20	0.17	0.15	0.14

COEFFICIENTS OF UTILIZATION

Luminaire	Distribution	Spacing not to exceed	RCR	80% Walls 50%	80% Walls 30%	80% Walls 10%	70% Walls 50%	70% Walls 30%	70% Walls 10%	50% Walls 50%	50% Walls 30%	50% Walls 10%
Category V — 2 T-12 Lamps – 430 MA, 1' wide prismatic wrap-around	7 up, 59 down	1.2 × mounting height	1	0.68	0.65	0.63	0.65	0.63	0.61	0.61	0.60	0.58
			2	0.60	0.56	0.53	0.58	0.55	0.52	0.55	0.52	0.49
			3	0.54	0.49	0.45	0.52	0.48	0.45	0.50	0.46	0.43
			4	0.49	0.43	0.40	0.47	0.43	0.39	0.45	0.41	0.38
			5	0.44	0.38	0.34	0.43	0.38	0.34	0.40	0.36	0.33
			6	0.40	0.34	0.30	0.39	0.34	0.30	0.37	0.32	0.29
			7	0.36	0.31	0.27	0.35	0.30	0.26	0.33	0.29	0.26
			8	0.32	0.27	0.24	0.32	0.27	0.23	0.30	0.26	0.23
			9	0.29	0.24	0.21	0.29	0.24	0.20	0.27	0.23	0.20
			10	0.27	0.22	0.18	0.26	0.21	0.18	0.25	0.21	0.18
Category V — 4 T-12 Lamps – 430 MA, 2' wide prismatic wrap-around	4 up, 59 down	1.3 × mounting height	1	0.66	0.64	0.61	0.64	0.62	0.60	0.61	0.59	0.57
			2	0.59	0.55	0.52	0.57	0.54	0.51	0.55	0.52	0.49
			3	0.53	0.48	0.45	0.52	0.48	0.44	0.49	0.46	0.43
			4	0.48	0.43	0.39	0.47	0.42	0.39	0.45	0.41	0.38
			5	0.43	0.38	0.34	0.42	0.37	0.34	0.40	0.36	0.33
			6	0.39	0.34	0.30	0.38	0.34	0.30	0.36	0.32	0.29
			7	0.35	0.30	0.26	0.34	0.30	0.26	0.33	0.29	0.26
			8	0.32	0.27	0.23	0.31	0.26	0.23	0.30	0.26	0.23
			9	0.28	0.24	0.20	0.28	0.23	0.20	0.27	0.23	0.20
			10	0.26	0.21	0.18	0.25	0.21	0.18	0.25	0.20	0.17
Category I — 2 Lamp strip – any loading	17 up, 69 down	1.6 × mounting height	1	0.83	0.79	0.75	0.79	0.76	0.72	0.73	0.70	0.67
			2	0.71	0.65	0.60	0.68	0.62	0.57	0.62	0.58	0.54
			3	0.62	0.55	0.49	0.59	0.53	0.47	0.55	0.49	0.44
			4	0.55	0.47	0.41	0.52	0.45	0.39	0.48	0.42	0.37
			5	0.48	0.40	0.34	0.46	0.38	0.33	0.42	0.36	0.31
			6	0.43	0.35	0.29	0.41	0.33	0.28	0.38	0.31	0.26
			7	0.38	0.30	0.25	0.36	0.29	0.24	0.34	0.27	0.23
			8	0.34	0.26	0.21	0.33	0.25	0.21	0.30	0.24	0.19
			9	0.30	0.23	0.18	0.30	0.23	0.18	0.27	0.21	0.17
			10	0.28	0.21	0.16	0.27	0.20	0.15	0.25	0.19	0.15
Category V — 1 Lamp – any loading, 2' wide, 1' deep prismatic lens	0 up, 60 down	1.2 × mounting height	1	0.64	0.62	0.60	0.63	0.61	0.59	0.60	0.59	0.57
			2	0.58	0.55	0.52	0.57	0.54	0.51	0.55	0.52	0.50
			3	0.52	0.48	0.45	0.51	0.47	0.44	0.49	0.46	0.44
			4	0.47	0.42	0.39	0.46	0.42	0.39	0.45	0.41	0.38
			5	0.42	0.37	0.34	0.42	0.37	0.34	0.40	0.36	0.34
			6	0.38	0.33	0.30	0.38	0.33	0.30	0.37	0.32	0.30
			7	0.35	0.30	0.26	0.34	0.30	0.26	0.33	0.29	0.26
			8	0.31	0.26	0.23	0.31	0.26	0.23	0.30	0.26	0.23
			9	0.28	0.23	0.20	0.28	0.23	0.20	0.27	0.23	0.20
			10	0.26	0.21	0.18	0.25	0.21	0.18	0.25	0.21	0.18
Category VI — 2 Lamp – any loading opaque sides	75 up, 5 down	1.5 × mounting height	1	0.68	0.65	0.62	0.59	0.56	0.54	0.42	0.41	0.39
			2	0.59	0.54	0.51	0.51	0.48	0.44	0.37	0.35	0.32
			3	0.52	0.46	0.42	0.45	0.40	0.37	0.32	0.29	0.27
			4	0.46	0.40	0.35	0.40	0.35	0.31	0.28	0.25	0.23
			5	0.40	0.34	0.30	0.35	0.30	0.26	0.25	0.22	0.20
			6	0.36	0.30	0.26	0.31	0.27	0.23	0.22	0.20	0.17
			7	0.32	0.26	0.22	0.28	0.23	0.19	0.20	0.17	0.14
			8	0.29	0.23	0.19	0.25	0.20	0.17	0.18	0.15	0.13
			9	0.26	0.20	0.17	0.23	0.18	0.15	0.17	0.13	0.11
			10	0.24	0.18	0.15	0.21	0.16	0.13	0.15	0.12	0.10
Category VI — Luminous ceiling – 50% transmission, 80% cavity reflectance	0 up, 69* down	1.5 to 2.0 × mounting height above diffuser	1	① For cavities that are painted white use 70% effective ceiling cavity reflectance. ② For cavities that are obstructed or have lower reflectances use 50% effective ceiling cavity reflectance.			0.60	0.58	0.56	0.58	0.56	0.54
			2				0.53	0.49	0.45	0.51	0.47	0.43
			3				0.47	0.42	0.37	0.45	0.41	0.36
			4				0.41	0.36	0.32	0.39	0.35	0.31
			5				0.37	0.31	0.27	0.35	0.30	0.26
			6				0.33	0.27	0.23	0.31	0.26	0.23
			7				0.29	0.24	0.20	0.28	0.23	0.20
			8				0.26	0.21	0.18	0.25	0.20	0.17
			9				0.23	0.19	0.15	0.23	0.18	0.15
			10				0.21	0.17	0.13	0.21	0.16	0.13
Category VI — Cove without reflector		Cove 12 to 18 inches below ceiling. Reflectors with fluorescent lamps increase coefficients of utilization 5 to 10%	1	0.42	0.40	0.39	0.36	0.35	0.33	0.25	0.24	0.23
			2	0.37	0.34	0.32	0.32	0.29	0.27	0.22	0.20	0.19
			3	0.32	0.29	0.26	0.28	0.25	0.23	0.19	0.17	0.16
			4	0.29	0.25	0.22	0.25	0.22	0.19	0.17	0.15	0.13
			5	0.25	0.21	0.18	0.22	0.19	0.16	0.15	0.13	0.11
			6	0.23	0.19	0.16	0.20	0.16	0.14	0.14	0.12	0.10
			7	0.20	0.17	0.14	0.17	0.14	0.12	0.12	0.10	0.09
			8	0.18	0.15	0.12	0.16	0.13	0.10	0.11	0.09	0.08
			9	0.17	0.13	0.10	0.15	0.11	0.09	0.10	0.08	0.07
			10	0.15	0.12	0.09	0.13	0.10	0.08	0.09	0.07	0.06

Ceiling cavity reflectances: 80%, 70%, 50%. Walls: 50%, 30%, 10%.

Index

340